# The Cassava Transformation

# The Cassava Transformation

## Africa's Best-Kept Secret

Felix I. Nweke
Dunstan S. C. Spencer
John K. Lynam

Michigan State University Press • *East Lansing*

♾ The paper used in this publication meets the minimum requirements
of ANSI/NISO Z39.48-1992 (R 1997) (Permanence of Paper).

Michigan State University Press
East Lansing, Michigan 48823-5202

Printed and bound in the United States of America.

07 06 05 04 03 02 1 2 3 4 5 6 7 8 9 10

LIBRARY OF CONGRESS CATALOGING-IN-PUBLICATION DATA
Nweke, Felix I.
The cassava transformation : Africa's best-kept secret / Felix I. Nweke,
Dunstan S. C. Spencer, and John K. Lynam
p. cm.
Includes bibliographical references and index.
ISBN 0-87013-602-X (alk. paper)
1. COSCA (Project) 2. Cassava–Africa. I. Spencer, D. S. C. III II. Lynam, John K.
. Title.
SB211.C3 N84 2002
633.6'82'096–dc21
2001005255

Cover and book design by Sharp Des!gns, Inc., Lansing, MI

Publication of *The Cassava Transformation: Africa's Best-Kept Secret* was made possible
by support from the Rockefeller Foundation and The International Institute for Tropical Agriculture.

Visit Michigan State University Press on the World Wide Web at:
www.msupress.msu.edu

# Dedication

*To*

**Professor Mohammed T. Dahniya**—an internationally acclaimed cassava breeder; Director of the Institute of Agricultural Research, Njala, Sierra Leone, from 1988 to 1999; member of the advisory committee of the Collaborative Study of Cassava in Africa (COSCA) from 1989 to 1999; and president of the International Society for Tropical Root Crops–African Branch (ISTRC-AB) from 1992 to 1998—who was murdered by rebels in Sierra Leone in January 1999

*and*

**Dr. Muamba Tshiunza**, a COSCA researcher from the Congo, who sustained a serious physical injury as a result of an automobile accident in Nigeria.

# Contents

# Figures

# Tables

# Preface

The people of Africa are bearing the brunt of food insecurity, malnutrition, and poverty. Cassava (manioc or tapioca) was introduced into Africa from South America in the sixteenth century and quickly spread throughout the continent. Today, cassava is the second most important food in the African diet. In the late 1950s, when African nations started to regain their independence, they were self-sufficient in food. Yet, in the early 1970s Africa became a net food importer. Because of poverty and a lack of foreign exchange, the countries of Africa are unable to import enough food to feed their growing populations. Therefore, the urgent challenge is to increase their food production.

Cassava can be a powerful poverty fighter if research-driven productivity gains increase production and drive down prices to rural and urban consumers. Cassava's adaptability to marginal soils and erratic rainfall conditions, as well as its high-yield per unit of land, make it an ideal crop in many farming systems in Africa. However, because of several long-standing myths and half-truths, cassava has been neglected and marginalized by many policymakers and by most donor and international agencies. Cassava's neglect in food policy circles is illustrated by the fact that there are only two books on cassava in Africa: *Manioc in Africa* by W. O. Jones (1959) and *Cassava in Shifting Cultivation* by L. O. Fresco (1986).

In the early 1980s, John Lynam, James Cock, and others at Centro Internacional de Agricultura Tropical (CIAT) in Colombia completed a study of cassava in Asia and South America and recommended that a similar study be carried out in sub-Saharan Africa. Dunstan Spencer of the International Institute of Tropical Agriculture (IITA), Ibadan, Nigeria, and Lynam and Cock prepared a proposal for a Collaborative Study of Cassava in Africa (COSCA). In the late 1980s, the COSCA study was initiated in six African countries and funded by the Rockefeller Foundation and the IITA. The study was based at the IITA and directed by Felix Nweke. During an eight-year period (1989 to 1997), COSCA researchers collected information on the cassava food system in 1,686 households in 281 villages in six countries: the Congo, the Côte d'Ivoire, Ghana, Nigeria, Tanzania, and Uganda. The COSCA data were analyzed by a team of analysts under the direction of Sunday Folayan at the IITA, Ibadan.

The COSCA study has unveiled one of Africa's best-kept secrets: the Cassava Transformation. This study describes how long-term research led to the development of high-yielding TMS varieties that have increased cassava yields by 40 percent without the addition of fertilizer. The new TMS cassava varieties have transformed cassava from a low-yield famine-reserve crop to a high-yield food staple, which is increasingly being converted into *gari* (dry cereal-like product) and consumed by rural and urban people.

We would like to thank the 1,686 farm families in the Congo, the Côte d'Ivoire, Ghana, Nigeria, Tanzania, and Uganda who provided comprehensive information on the role of cassava in their farming systems and households. In-depth studies were carried out by dedicated national research teams in the six countries. Two Directors General of the IITA, the late Lawrence Stifel and Lukas Brader provided invaluable support, encouragement, and wise counsel. We would also like to thank John Strauss, S. K. Hahn, F. M. Quin, John Thackway, Doyle Baker, Jacques

Eckebil, Hans Herren, Peter Neuenschwander, Dyno Keatinge, Y. Jeon, L. Halos Kim, Hiroki Inazumi, Randolph Barker, Robert Cunningham, Rupert Best, James Cock, David Norman, Derek Byerlee, Joseph Mukiibi, O. A. Edache, L. S. Ene, Francis Ofori, André-Joseph Onyembe, Dannis Ekpete, Patrick Ngoddy, Maxwell Ikeme, Frank Ndili, Donatus Ibe, Fred Ezedinma, Walter Enwezor, Peter Ay, Rose Umelo, and Chris Bankole.

The International Society for Tropical Root Crops-African Branch (ISTRC-AB) has been very supportive of the COSCA study. Special thanks are extended to the following members of the Society: Malachy Akoroda, Robert Asiedu, Mpoko Bokanga, Mohammed Dahniya, Alfred Dixon, Regina Kapinga, William Khizza, Nzola Mahungu, Chikelu Mba, H. D. Mignouna, Adeyinka Onabolu, George Otim-Nape, John Otoo, and James Whyte. The collaborating individuals and agencies are listed in appendix 2.

The field research for chapter 12 on the new uses of cassava was financed by the FAO. Special thanks are due to Louise Fresco, Morton Satin, David Wilcock, Marcio Porto, NeBanbi Lutaladio, and Rosa Rolle. Thanks are also due to Afuekwe Nweke, Tunde Adegbesan, Samuel Ifeacho, Anthony Ezekwesili, Gibson Okeke, Paul Okeke, Augustine Esedo, Jonas Lemchi, and Godwin Asumugha who assisted in the field research.

After the COSCA field studies were completed in 1997, Felix Nweke prepared the draft of the book manuscript while he was a Visiting Professor in the Department of Agricultural Economics at Michigan State University. Dr. Larry Hamm, Chairperson; Ann Robinson, Allan Schmid, Chris Wolf, Brian Hoort, Linda Beck, Thomas Reardon, Eric Crawford, Julie Howard, Thomas Jayne, Cynthia Donovan, and other members of the Department of Agricultural Economics offered valuable administrative and technical support and encouragement. Special thanks are also extended to the members of the Bailey Scholars Program and African Studies Center at Michigan State University who offered valuable support and encouragement. Felix Nweke's residence at Michigan State was funded by the Onyeyili-Nweke Fund for a Better Africa and the Rockefeller Foundation. S. K. Hahn, Carl Eicher, John Strauss, Glenn Johnson, Obinani Okoli, Bede Okigbo, Eric Tollens, Gregory Scott, and Nwogo Nweke provided incisive comments on several drafts of the manuscript.

CHAPTER 1

# Cassava and Africa's Food Crisis

## Introduction

Sub-Saharan Africa (hereafter Africa) is a continent in crisis; it is racked with hunger, poverty, and the HIV/AIDS pandemic. Africa is also the region with the fastest population growth, the most fragile natural resource base, and the weakest set of agricultural research and extension institutions. However, when African nations started to regain their independence in the late 1950s and early 1960s, they were self-sufficient in food production and leading exporters of cocoa, coffee, rubber, sisal, groundnuts (peanuts), and palm oil, respectively. By contrast, Asia was the epicenter of the world's food crisis in the 1960s and 1970s.

Africa became a net food importer in the early 1970s, and food production grew at half of the population growth rate from 1970 to 1985. Today, the people of Africa are bearing the brunt of world food insecurity, malnutrition, and poverty. For example, a recent International Food Policy Research Institute (IFPRI) study points out that child malnutrition is expected to decline in all major developing regions except Africa, where the number of malnourished children is forecast to increase by about 30 percent by 2020 (Pinstrup-Anderson et al. 1999, 6).

The average GNP per capita in Africa in the year 2000 was US$480 (World Bank 2000). Many countries have been destabilized by civil strife and authoritarian regimes. Approximately 75 percent of the poor in Africa are rural people who secure their livelihood from farming and livestock or from nonfarm activities that depend mostly on agriculture.

Africa's population is expected to double to 1.2 billion by 2020, and its urban population will likely grow at an even faster rate. With urbanization and higher incomes, the composition and characteristics of food demand will be significantly altered. More of the food supply will have to be processed, transported, and stored (McCalla 1999).

Without question, domestic food production and food imports will have to be increased to meet Africa's growing food demand. Yet because of poverty and a lack of foreign exchange, Africa's net cereal imports are expected to remain low (Pinstrup-Anderson et al. 1999). Therefore, the urgent challenge before African nations is to increase domestic food production.

Because of the success of the Green Revolution in Asia, many "instant experts" on Africa have raised expectations about replicating the Asian Green Revolution model in Africa. However, Asia's Green Revolution was based on high-yielding rice and wheat varieties grown under irrigation. Yet irrigation presently involves only about 5 percent of the land under cultivation in Africa, and wheat cannot be grown in the rainforest and semi-arid areas of Africa. Because of the sharp differences in the agroecologies between Asia and Africa and the larger number of food staples in Africa, attempts to promote the Asian Green Revolution in Africa in the 1970s and 1980s have turned out to be embarrassing failures. The failure of Asian Green Revolution models to take root in Africa has spurred donors to increase their support for strengthening Africa's national agricultural research systems (NARs) (Byerlee and Alex 1998) and to intensify research on Africa's major food staples: maize, cassava, sorghum, rice, wheat, millet, yam, and banana.[1]

Maize is Africa's most important food crop, and it has been studied extensively and is held up as a model food crop to meet Africa's growing urban demand for convenient food products (Mellor, Delgado, and Blackie

1987; Blackie 1990; Byerlee and Eicher 1997). Yet maize production is risky because of undependable rainfall, and development of irrigation for maize production is not technically and financially feasible for most African countries. With this background, we pose the question: why cassava?

## Why Cassava?

The dramatic cassava transformation that is under way in Nigeria and Ghana is Africa's best-kept secret. This transformation describes how the new Tropical Manioc Selection (TMS) varieties have changed cassava from a low-yielding famine-reserve crop to a high-yielding cash crop that is prepared and consumed as a dry cereal (*gari*).[2] This book focuses on the cassava transformation for two important reasons. First, cassava is Africa's second-most important food staple in terms of per capita calories consumed. Cassava is a major source of calories for roughly two out of every five Africans. In some countries, cassava is consumed daily, and sometimes more than once per day. In the Congo, cassava contributes more than one thousand calories per person per day to the average diet, and many families eat cassava for breakfast, lunch, and dinner. Cassava is consumed with a sauce made with ingredients rich in protein, vitamins, and minerals. In the Congo and Tanzania, cassava leaves are consumed as a vegetable. Cassava leaves are rich in protein, vitamins, and minerals.

Second, the literature on cassava, however, is out of date and laden with long-standing myths and half-truths. For example, only two books have been published on cassava in Africa in the past forty years: *Manioc in Africa* by W. O. Jones (1959) and *Cassava in Shifting Cultivation* by L. O. Fresco (1986).[3] Today, cassava is a marginalized crop in food policy debates because it is burdened with the stigma of being an inferior, low-protein food that is uncompetitive with such glamour crops as imported rice and wheat. Many food policy analysts consider cassava an inferior food because it is assumed that its per capita consumption will decline with increasing per capita incomes.

This book describes how the new TMS cassava varieties have transformed cassava from a low-yielding famine-reserve crop to a high-yielding cash crop. With the aid of mechanical graters to prepare *gari,* cassava is increasingly being produced and processed as a cash crop for urban consumption in Nigeria and Ghana. This book unveils the secret of the cassava transformation and points out the potential of cassava for feeding Africa, helping the poor, and generating foreign exchange from the sale of cassava as an industrial raw material and as livestock feed.

This book addresses three audiences. First, it presents comprehensive information on the cassava industry in six African countries to help enlighten African policy makers on the importance of investing in increasing the productivity of the cassava industry and driving down the real (inflation-adjusted) cost of cassava to rural and urban consumers. Second, the book is aimed at researchers working on cassava, including specialists in biochemistry, plant breeding, agronomy, farm mechanization, processing, and nutrition. Third, the book is designed to inform members of international and donor agencies and Non-Governmental Organizations (NGOs) that cassava has the potential to increase farm incomes, reduce rural and urban poverty, and help close the food gap. Without question, cassava holds great promise for feeding Africa's growing population. With the aid of the new TMS varieties, smallholders (small-scale family farms) are getting 40 percent higher yields without fertilizer.

We shall show in subsequent chapters that cassava can be produced with family labor, land, and a hoe and machete, making it an attractive and low-risk crop for poor farmers. Also, cassava is available to low-income rural households in the form of simple food products (for example, dried roots and fresh leaves), which are significantly cheaper than grains such as rice, maize, and wheat. In addition, urban households in many parts of West Africa consume cassava in the form of *gari* (granulated cassava), a convenient food product.

Cassava has several other advantages over rice, maize, and other grains as a food staple in areas where there is a degraded resource base, uncertain rainfall, and weak market infrastructure. Whereas cassava can be grown under a wide range of ecological conditions, rice has a narrow ecological adaptation in Africa. Although upland rice varieties are currently being developed, the best available varieties, in terms of yield and grain quality, are wetland varieties. Yet wetland (irrigated or flooded) areas are limited in Africa, and most such lands are expensive to develop (Federal Agricultural Coordinating Unit 1986).

Cassava is drought tolerant, and this attribute makes it the most suitable food crop during periods of drought and famine. Cassava has historically played an important famine-reserve role in Eastern and Southern Africa, where maize is the preferred food staple and drought is a recurrent problem. In the 1920s and 1930s, the British colonial authorities were concerned about the frequent famines in the grain-consuming areas of Eastern and Southern Africa, which were caused by drought and locusts. The colonial governments encouraged local farmers to plant cassava as a famine-reserve crop and established cassava research programs in Tanzania (formally Tanganyika) and in other colonies to find a solution to the cassava mosaic virus disease. Cassava has now come to play a leading role in African agriculture, not only as a famine-reserve crop, but also as a cash crop for urban consumption.

## Cassava: Introduction and Diffusion in Africa

Cassava (*Manihot esculenta Cranz*), variously designated as manioc, mandioc, tapioca, or yucca is a perennial woody shrub of the *Euphorbiacae* family.[4] It is grown principally for its swollen roots but its leaves are also eaten in some parts of Africa. The roots are from 25 to 35 percent starch; the leaves, though unimportant as a source of calories, contain a significant amount of protein and other nutrients. Cassava is adapted to the zone within latitudes 30° north and south of the equator, at elevations of not more than two thousand meters above sea level, in temperatures ranging from 18°C to 25°C; to rainfall of fifty to five thousand millimeters annually; and to poor soils with a pH from 4 to 9 (fig. 1.1).

Cassava was introduced into Africa by Portuguese traders from Brazil in the sixteenth century. A native of South America and Southern and Western Mexico, cassava was one of the first crops to be domesticated. There is archeological evidence that it was grown in Peru four thousand years ago and in Mexico some two thousand years ago (Okigbo 1980).

**Figure 1.1. Areas of Cassava Production in Africa.** *Source: Okigbo 1980.*

From Mid- and South America, cassava spread to the West Coast of Africa and the Congo in the late sixteenth century, probably in slave ships. The technique of making *gari* from cassava roots was introduced in Sao Tome about 1780, a discovery that aided in the diffusion of cassava in West Africa. Cassava was introduced into East Africa (Madagascar and Zanzibar) via Reunion by the end of the eighteenth century. It was widely grown in Africa and Southeast Asia by the 1850s (Okigbo 1980).

The diffusion of cassava can be described as a self-spreading innovation in African agriculture. It was initially adopted about four hundred years ago as a famine-reserve crop. In the Congo, where the crop was first introduced, millet, banana, and yam were the traditional staples but farmers adopted cassava because it provided a more reliable source of food during drought, locust attack, and the "hungry season."[5] Although there was some local trade in cassava, production was mostly for home consumption and cassava was prepared in the simplest fashion, that is, by slicing and boiling (Jones 1959).

Currently, about half of the world production of cassava comes from Africa. Cassava is cultivated in about forty African countries, stretching through a wide belt from Madagascar in the Southeast to Senegal and Cape Verde in the Northwest. Around 70 percent of Africa's cassava output is harvested in Nigeria, the Democratic Republic of Congo (hereafter the Congo), and Tanzania (International Fund for Agricultural Development and Food and Agriculture Organization 2000). Throughout the forest and transition zones of Africa, cassava is either a primary or a secondary food staple.

The reasons for the rapid spread of cassava cultivation in Africa include the following:

- It adapts to poor soils on which many other crops fail.
- It is easily propagated by stem cuttings.
- It resists droughts, except at planting time, and it resists locust damage, making it a good famine-reserve crop.
- It has a relatively high yield and is a low-cost source of calories. It can produce more carbohydrates per hectare than any other food staple and can be harvested as needed.
- It can be planted at any time of the year, provided there is enough moisture for stem cuttings to take root. Cassava roots can be left in the ground and harvested from six to forty-eight months from planting (Okigbo 1980).

**Table 1.1.** The Four Stages of Cassava Transformation.

<table>
<tr><th></th><th>I. Famine Reserve Crop</th><th>II. Rural Food Staple</th></tr>
<tr><td>DRIVING FACTORS</td><td>Tanzania</td><td>Congo, Côte d'Ivoire, and Uganda</td></tr>
<tr><td>CASSAVA PRODUCTION OBJECTIVE</td><td>• Mostly for home consumption<br>• Cassava is a secondary food staple</td><td>• Mostly for home consumption<br>• Cassava is the primary food staple in the Congo<br>• Cassava is a secondary food staple in Côte d'Ivoire and Uganda<br>• Cassava is a family food staple in households producing tree crops, such as cocoa in Côte d'Ivoire and coffee in Uganda</td></tr>
<tr><td>POLITICAL AND ECONOMIC ENVIRONMENT</td><td>• In areas with uncertain rainfall:<br>–Colonial governments encouraged cassava production as a famine-reserve crop starting in the 1920s<br>–Since independence, Tanzanian government encouraged cassava production during droughts</td><td>• The governments of the Congo and Côte d'Ivoire encourage the importation of rice and wheat</td></tr>
<tr><td>TECHNOLOGY DEVELOPMENT</td><td colspan="2"></td></tr>
<tr><td>1. Genetic Improvement</td><td colspan="2">• Colonial governments established research stations in Tanzania, Nigeria, the Congo, etc.<br>• Research priorities focused on controlling cassava mosaic virus and brown streak virus</td></tr>
<tr><td>2. Seed</td><td colspan="2">• Farmer-to-farmer exchange of planting materials</td></tr>
<tr><td>3. Agronomic Practices</td><td colspan="2">• Farmers planted at will, compatible with labor demand schedule for cash crop</td></tr>
<tr><td>4. Weeding</td><td colspan="2">• Occasional weeding using hand hoe by family labor</td></tr>
<tr><td>5. Harvesting</td><td colspan="2">• Partial harvesting with hand hoe, using family labor</td></tr>
<tr><td>6. Processing</td><td colspan="2">• Manual processing with hand tools</td></tr>
<tr><td>7. Food Products</td><td colspan="2">• Roots eaten in fresh form, or as pastes, as well as dried<br>• Leaves for food</td></tr>
</table>

| | III. Cash Crop for Urban Consumption | IV. Livestock Feed & Industrial Raw Material |
|---|---|---|
| | Nigeria and Ghana | ---- |
| | • Mostly for sale as *gari* in urban centers | • Industrial starch and pellets for export |
| | • Increasing urban demand for convenient foods<br>• Improved rural roads for easy farmer access to market centers<br>• Government policies encourage the replacement of imported wheat and rice with cassava food products | • Policy of substitution of cassava starch for imported starch<br>• Export promotion |
| | • IITA developed high-yielding TMS varieties<br>• IITA organized training programs for national cassava scientists | • Early (under 12 months) bulking varieties<br>• Cassava roots suitable for mechanical harvesting and peeling |
| | • National research and extension programs and private-sector agencies multiply and distribute planting materials of improved varities | • Private seed companies |
| | • Timely planting<br>• High stand density<br>• Low frequency of intercropping | • Timely planting<br>• Optimum stand density<br>• Mono-cropping |
| | • Regular weeding with hand hoe by hired labor | • Regular weeding<br>• Mechanized weeding |
| | • Complete harvesting<br>• Harvesting at 12 months or less<br>• Harvesting with hand hoe<br>• Use of hired labor | • Harvesting in 12 months or less<br>• Mechanized harvesting |
| | • Partly manual, partly mechanized | • Fully mechanized |
| | • Convenient food products | ---- |

## The Cassava Transformation

Traditionally, cassava is produced on small-scale family farms. The roots are processed and prepared as a subsistence crop for home consumption and for sale in village markets and shipment to urban centers. Over the past thirty to fifty years, smallholders in Nigeria and Ghana have increased the production of cassava as a cash crop, primarily for urban markets. This shift from production for home consumption to commercial production for urban consumers, livestock feed, and industrial uses can be described as the cassava transformation. During this transformation, high-yielding cassava varieties have been developed to increase yields while labor-saving and improved-processing technologies have reduced the cost of producing and processing cassava food products to the point where they are competitive with food grains such as wheat, sorghum, and rice for urban consumers. Looking ahead, as the costs of cassava production, harvesting, processing, and marketing continue to be reduced, we can expect cassava to play an expanded role as a source of industrial raw material in Africa and as a source of foreign exchange earnings through the export of cassava pellets for livestock feed.

The cassava transformation encompasses four stages: Famine Reserve, Rural Food Staple, Urban Food Staple, Livestock Feed, and Industrial Raw Material (table 1.1).

### ■ Stage I: Famine Reserve

Today in many countries in Eastern and Southern Africa, maize is the preferred food staple, while cassava is regarded as a famine-reserve crop. If there is a bumper grain crop, cassava is often left unharvested in the ground for up to four years. In the famine-reserve stage, cassava yields are low, about 10 tons per hectare, and the plant is usually harvested late, and often on a meal-to-meal basis. In some countries, such as Tanzania and the Congo, cassava leaves are harvested and consumed as a vegetable. In Tanzania, for example, the ministry of agriculture usually organizes crash cassava production programs when the maize crop is threatened by drought. Yet, after the drought is over, most governments curtail these special extension programs. As a result, there is little continuity in research and extension, and cassava farmers are typically forced to rely on a farmer-to-farmer exchange of varieties, especially those that extend the storage life of cassava in the ground.

### ■ Stage II: Rural Food Staple

In the rural food staple stage, cassava is the main source of calories in the diets of rural consumers, and farmers plant local varieties with low genetic potential and achieve low yields. Production, harvesting, and processing tasks are carried out manually, and farm households consume most of the output. The Congo is currently in the rural food staple stage because poor roads, grinding poverty, and political chaos have kept the rural people locked into a virtual subsistence agriculture. Cassava is consumed mostly as dried roots, and cassava leaves are the main vegetable in rural diets. In most of the Côte d'Ivoire and Uganda, where tree crops such as cocoa and coffee are grown, farmers grow cassava as their main food staple because tree crop production requires peak labor inputs, mainly at planting and harvesting while cassava production does not require seasonal labor. Cassava roots are boiled and eaten because sun-drying of cassava roots is an inefficient method in the forest zone.

### ■ Stage III: Urban Food Staple

An African nation is in the urban food staple stage of the cassava transformation when cassava is produced primarily as a cash crop and processed mostly for sale in urban markets. Both Nigeria and Ghana are in the urban food staple stage of the transformation. In these countries commercial production and processing of cassava products for urban markets is driven by high-yielding cassava varieties, increasing urban demand for food, availability of improved rural roads for the transport of cassava to urban market centers, and government policies that encourage the substitution of cassava products for imported rice and wheat. In Nigeria, for example, high-yielding TMS cassava varieties have increased on-farm yields by 40 percent without fertilizer. These varieties are resistant to common cassava pests and diseases and they have helped increase farm incomes and reduce rural and urban poverty.

During the urban food staple stage, cassava is produced and processed into a variety of low-cost, convenient food products for sale in urban centers and foreign markets. The technological requirements for a nation to move to the urban food staple stage include high-yielding and early-bulking cassava varieties that can be harvested at fewer than twelve months; mechanization of harvesting and some processing tasks to improve labor productivity; and the development of an array of convenient cassava food products, such as cassava chips and food snacks.[6] During this stage, private firms assume a greater role in supplying to farmers planting materials; mechanized services for processing tasks; and marketing services. Private food processing firms also undertake research and development activities leading to new cassava food products.

### ■ Stage IV: Livestock Feed and Industrial Raw Material

The cassava industry advances to the livestock feed and industrial raw material stage in countries where the production, processing, and marketing costs are reduced enough to enable African cassava to compete in global industrial starch and livestock feed markets. However, cassava has not advanced to the livestock feed and industrial raw material stage in any African country. The preconditions for a country to advance to the livestock feed and industrial raw material stage call for labor-saving production and processing technologies, early-bulking varieties that are suitable for mechanized harvesting and peeling, and improved drying facilities. An efficient and well-integrated production and marketing system is also necessary, to assure a steady supply of cassava products to domestic industries and European markets. Public and private investments in research and development are also required to develop cassava products for industrial uses. Private sector initiative is required to supply planting materials and processing and marketing services.

This discussion has traced the evolution of the cassava industry in Africa through the first three stages—famine reserve, rural food staple, and urban food staple—of the cassava transformation. Presently, Tanzania is at the first stage of the cassava transformation; the Congo, the Côte d'Ivoire, and Uganda are at the second stage; and Nigeria and Ghana are in the third stage of the transformation. As previously mentioned, there are currently no African nations in the fourth stage of the transformation.

## The Collaborative Study of Cassava in Africa (COSCA)

African policy makers and members of the international donor agencies and NGOs are searching for a solution to the continent's food production crisis. Yet basic information is lacking on cassava's growing conditions and the economics of production, processing, and marketing. There is also a dearth of information on market opportunities for expanding the use of cassava in industrial markets and for livestock feed in Africa and in Europe.

Many of the available production statistics are, at best, educated guesses. Thus, policy makers and donor agencies are left in doubt about the crucial importance of cassava in the African food economies, while plant-breeding research on the crop is being carried out with little knowledge of the agronomic, processing, and marketing problems faced by the small-scale farmers and traders.

A major study of cassava, the Collaborative Study of Cassava in Africa (hereafter the COSCA study) was carried out from 1989 to 1997 under the aegis of the International Institute of Tropical Agriculture (IITA) in Ibadan, Nigeria. The study was financed by the Rockefeller Foundation and the IITA.

Over the period from 1989 to 1995, COSCA researchers collected primary data in six African countries where roughly 70 percent of the total cassava production in Africa takes place: the Congo, the Côte d'Ivoire, Ghana, Nigeria, Tanzania, and Uganda. These countries will hereafter be referred to as the six COSCA study countries. The data collected included information on cassava production systems, processing and food preparation methods, market prospects, and consumption patterns. This information will be used to help develop improved food policies and research and extension programs in order to accelerate the cassava transformation and ultimately increase food security and incomes of the people of Africa.

The COSCA research team collected information in the six study countries at the village, household, and field levels as follows:

- *Level I.* Characterization of cassava-producing areas at the village level:
  - Environment (physical, social, economic)
  - Production
  - Processing
  - Marketing
  - Consumption
- *Level II.* Cassava production at the field level:
  - Land area
  - Production practices
  - Input/output
  - Yield
  - Utilization (sale/home use, processing/fresh use)
- *Level III.* Detailed studies on post-harvest issues at the household and market levels:
  - Processing
  - Marketing
  - Consumption

Comprehensive data covering Levels I, II, and III were collected from 281 villages in the six countries under study. The locations of these villages are shown in figure 1.2. The methods of the study are presented in appendix 1.

The information provided in this book is taken from the six COSCA study countries, supplemented with secondary data from the Food and Agriculture Organization of the United Nations (FAO) and information collected from African farmers, researchers, traders, and processors throughout Africa from the beginning of the COSCA study in 1989 to 2001.

**Figure 1.2.** Location of Survey Sites. *Source: COSCA Study.*

## Plan of This Book

Chapter 2 is devoted to a discussion of five common myths and half-truths that have contributed to the low priority that cassava has been given in African research and food production programs. These myths and half-truths help explain why cassava has received only token and periodic support from donors. Evidence is marshaled to show that these myths and half-truths are based on misconceptions and out-of-date information.

Chapter 3 pulls together data from Africa, Asia, and Latin America to show the five important roles that cassava can play in African development: famine-reserve crop, rural food staple, cash crop for urban consumption, industrial raw material (e.g., cassava starch), and earner of foreign exchange through the export of cassava pellets for livestock feed in Africa and Europe. Chapter 4 describes cassava production, highlighting the factors that have influenced its historical development in Africa.

Chapter 5 deals with the historical evolution of genetic research on cassava in Africa through the joint efforts of farmers, the private sector, national agricultural research systems (NARs), and the International Agricultural Research Centers (IARCs). Special attention is devoted to tracing the fascinating history of cassava breeding in Africa, including the story of how IITA cassava breeders drew on varieties resistant to mosaic virus that had been developed in the 1930s by British colonial researchers at the famous East African Agricultural and Forestry Research Station at Amani (hereafter the Amani research station) in Tanzania (formally Tanganyika). Under the distinguished leadership of Dr. Sang Ki Hahn at IITA in Ibadan, Nigeria, high-yielding TMS varieties were developed by crossing the Amani varieties with West African cassava varieties. Without doubt, the high-yielding TMS varieties have fueled the cassava transformation, especially in Nigeria and Ghana. The TMS varieties developed by Dr. S. K. Hahn and his colleagues have boosted on-farm yields by 40 percent without the addition of fertilizer.

Without question, the new TMS varieties can help Africa solve the continent's chronic food deficit, especially in countries where cassava is a food staple. But there is still a need for a great deal of public and private research and investment in developing improved harvesting and food preparation technologies. Research is also needed in new cassava food products as well as improved types of cassava starch to compete with maize, rice, and other types of starch for industrial uses such as paint and plastic.

Chapter 6 is devoted to a discussion of COSCA findings on agronomic field practices such as weeding, spacing, fallow management, and the like, and how they are modified over time by the farmers in response to changes in population pressure, market demand, and pests and diseases problems. Chapter 7 discusses the diffusion and profitability of the IITA's new TMS varieties and shows that the spread of these varieties has been accelerated with the introduction of labor-saving processing technology.

Chapters 8 and 9 discuss the traditional processing and food preparation practices and show how improved cassava processing technology has induced an expansion in cassava production. The two chapters call for the diffusion of improved *gari* preparation methods throughout cassava-producing countries in Africa. Chapter 10 challenges the half-truth that cassava is a women's crop by presenting evidence that as cassava is transformed into a commercial crop, both men and women play significant roles in production and processing. However, women play the dominant role in food preparation during each of the four stages of the cassava transformation.

Chapter 11 challenges the myth that cassava is an "inferior food" by presenting empirical evidence from rural household consumer surveys by the COSCA researchers in Nigeria, Tanzania, and Uganda and urban household consumer surveys by other researchers in Nigeria and Ghana. The results of these surveys reveal high-income elasticities of demand for cassava in rural and urban areas, and show that cassava is a food staple of low-, middle-, and high-income households. The results of the surveys also show that the type of cassava product consumed varies by income level. Low-income rural households, for example, consume cassava in its cheapest forms, such as dried roots, while middle- and high-income rural households and all types of urban households consume the crop in convenient food forms, such as *gari*.

Using Nigeria as a case study, chapter 12 examines what needs to be done to increase the use of cassava in livestock feed and as industrial raw material. Special attention is given to use of cassava starch in the textile, pharmaceutical, and petroleum drilling industries; use of cassava syrup concentrate in the soft drink industry; and use of dried cassava roots in the beer malt, alcohol, and ethanol industries.

Chapter 13 synthesizes the results of the eight-year COSCA study and shows that the cassava transformation that is under way in Nigeria and Ghana has turned cassava into a cash crop that has a large potential for meeting the food security needs of the rising urban population in Africa. Now special attention is needed to speed up the cassava transformation in other countries where cassava is either a primary or a secondary food staple.

CHAPTER 2

# Myths and Half-Truths

## Introduction

Forty years ago W. O. Jones (1959) reported that cassava was a controversial crop in Africa and the subject of heated disagreement among academic specialists. The advocates of cassava praised it because it produced the largest number of calories per hectare of any crop and could be grown on poor soils and withstand severe attacks of drought, pests, and diseases. These attributes explain why many colonial governments encouraged and, in some cases, forced smallholders to grow the crop.

At the same time, however, many critics pointed out that cassava often contains lethal quantities of cyanogens (prussic acid), which can kill humans and animals. The critics also noted that cassava is deficient in protein and that its high yields impoverish the soil. Finally, the critics contended that cassava was a "women's crop," which was produced and consumed by impoverished rural households. These criticisms explain why many African policy specialists since independence have been preoccupied with increasing the production of wheat and rice to feed Africa's urban population. In fact, the historical bias in favor of rice and wheat in food policy circles is palpable and disconcerting. In 1958, for example,

Johnston described rice as the "glamour crop" of West Africa (1958, 226). Later, Jones reported that African consumers described wheat flour as a "delicacy" (1972, 28).

Today, four decades after Jones's seminal book on cassava in Africa, the debate over cassava is still heated, and marked by many myths, half-truths, and misunderstandings. In this chapter, we critically discuss these debates and draw on COSCA's rich empirical database to "update" the cassava story to the present year.

## Cassava as a Subsistence Crop

Historically, with 90 to 95 percent of the people in Africa engaged in farming, cassava has been a subsistence crop. Today, although only one-half to two-thirds of the people of Africa live in rural areas, there is still a myth that cassava is primarily a subsistence crop, grown for home consumption by farmers and rural net food buyers.

The COSCA studies, however, challenge this myth and show that cassava is increasingly becoming a cash crop which smallholders sell to rural and urban consumers. The percentages of cassava fields planted for sale in the six COSCA countries were as follows: the Congo, 59; the Côte d'Ivoire, 51; Ghana, 57; Nigeria, 45; Tanzania, 32; and Uganda, 25. In the Congo, the proportion of cassava planted for sale was higher than for any other crop (Tollens 1992). In Nigeria, the Congo, and elsewhere, there are small-scale farms that produce five to ten hectares of cassava entirely for sale (Berry 1993).

An average of 40 percent of the sample villages in the six COSCA countries sold cassava in the fields before harvesting; 15 percent sold the crop at home after harvesting; and 45 percent sold it in village markets after harvesting. Thirty-five percent of the same villages sold fresh cassava roots to traders, 20 percent sold to processors, and 45 percent sold directly to consumers as their most frequent buyers. The traders and the processors resold cassava in urban markets.

In Nigeria, farmer access to production credit has led to an increase in cassava planting for the market. For example, in 1991 the COSCA farmers in Nigeria who secured credit for cassava production sold 56 percent of their total cassava production, while those who did not use credit sold only 40 percent. About 30 percent of the COSCA study farmers in Nigeria secured credit in 1991, all of it from informal credit sources such as the traditional moneylenders and the cassava traders. The traditional moneylenders charged excessively high interest rates. The cassava traders required, as a condition for their credit, that the cassava be sold to them (Ay 1991).

The COSCA data from 281 villages throughout Africa dismiss the myth that cassava is a subsistence crop produced by and for rural households, because cassava is increasingly being transformed into a cash crop that is produced and marketed as an urban food staple in many countries in Africa.

## Cassava as a Soil Nutrient-Depleting Crop

Historically, many scholars have asserted that cassava depletes the soil. For example, Hendershott et al. report that "cassava is well known not only for producing large quantities of carbohydrate, but also for exhausting the soil" (1972, 60). Similar assertions have been made by Davesne (1950), Irvine (1953), and Grace (1977). Soil fertility is a subject of major importance in a discussion of expanding food production in Africa (International Institute of Tropical Agriculture 1998). Human-induced land degradation is severe in Africa, with soil scientists reporting that 30 percent of the agricultural land, pastures, forests, and woodlands are degraded. Numerous researchers have claimed that a lack of replenishment of nutrients is leading to rapid deterioration in soil fertility. Fertilizer use is low because of high transport costs, late delivery, and the risks associated with food production in marginal areas (Pinstrup-Anderson et al. 1999).

To test the hypothesis that cassava depletes the soil, COSCA researchers collected soil samples from 1,501 fields planted to staple crops in 281 villages in the cassava-growing areas of the Congo, the Côte d'Ivoire, Ghana, Nigeria, Tanzania, and Uganda. The soil samples collected over the

1991 to 1992 period revealed that the amount of clay, silt, and sand in the soil did not differ between cassava fields and those of other crops.[1] In fact, soils of cassava fields were higher in total Nitrogen, organic matter, Calcium, Sodium, total exchangeable bases, effective cation exchange capacity, and pH.[2]

The myth of cassava being a soil nutrient-depleting crop may be attributed to the fact that cassava is widely grown in the forest zone, where high rainfall and sandy soil cause organic matter decomposition, leaching, and soil erosion to occur at a faster rate than in the transition and the savanna zones.[3] For example, an eight-year experiment in the forest zone of Ghana revealed that cassava yields declined under a continuous rotation of cassava and maize with fertilizer. In the transition zone, a similar experiment was carried out without the application of fertilizer, and cassava yields declined at a slower rate than in the forest zone (Nye and Greenland 1960, cited in Odurukwe and Oji 1981).

A twenty-year cassava yield experiment conducted by S. K. Hahn at the IITA's high rainfall station at Ibadan in the transition zone from the early 1970s until the early 1990s found that the yield dropped significantly from an average of 40 tons per hectare during the first four years and then stabilized at around 20 tons per hectare from the fifth to the twentieth year.[4] In this experiment the cassava was grown every other year in a two-year rotation without chemical fertilizer. Hahn concluded that cassava produces a large amount of foliage which is recycled as manure to the soil. Hahn's landmark twenty-year study shows that cassava yields are sustainable under continuation cultivation.[5] The COSCA soil survey and Hahn's twenty-year yield experiment provide evidence that the assertion that cassava depletes the soil is a half-truth.

The myth of cassava as a soil nutrient-depleting crop has been used to downgrade cassava's role as an environmentally friendly crop. Cassava cultivation entails minimal soil disturbance, especially in light soils, which are susceptible to wind or water erosion.[6] The plant provides soil cover so long as it grows; as a semiperennial it does not shed all its leaves with senescence. As a semiperennial, cassava plants serve as a planted fallow; although there are places, especially in high population density and market center areas, where cassava is harvested from six to twelve months after planting, the normal time of harvest is twelve months or more after planting. The cassava plant protects the soil by providing cover and recycles nutrients by shedding old leaves as it grows.

To summarize, the evidence on cassava and soil nutrient depletion is mixed. The COSCA soil studies in the six study countries reveal that soil nutrients are as high in cassava fields as in fields of other crops. Hahn's twenty-year cassava yield experiment shows that cassava yields decline for the first four years and then stabilize for the next sixteen years.

Yet the evidence on cassava and soil fertility in the forest zone shows that cassava yields declined (even with the application of fertilizer) over the eight-year period of the experiment. This decline in cassava yield is similar to decline in other food crop yields in the forest zone. In light of these findings, we can state that it is half-truth to assert that cassava ruins the soil.

## Cassava as a Women's Crop

In 1970, Esther Boserup published an influential book that dramatized the role of women in development. Boserup pointed out that women play the dominant role in food production in Africa. In 1975, Boserup's book formed the intellectual centerpiece for a United Nations global conference on women in development, held in Mexico City. That same year, the United Nations Economic Commission for Africa set up an African Training and Research Center for Women. The cofounder of the center, Margaret Snyder, proclaimed that women produced 80 percent of the food in Africa (1990, v). Yet Snyder's bold assertion was unencumbered with hard empirical evidence.

Buoyed by the work of Boserup and the enthusiasm following the Mexico City conference, many universities set up offices of Women in Development in the 1970s and 1980s. After several decades of research on farm households in Africa, however, it was found that the role of women in farming and processing can be better understood by studying the changing roles of both women and men in development over time. As a result, the policy debate shifted in the 1990s from women in development to gender and development. Looking back, it is obvious that the role of women in various farming activities can be best understood if the roles of men and children in these activities are also included in the analyses of decision making and labor allocation in farm households. This evolution in thought from women to gender issues is analogous to what Albert Hirschman concluded in his 1981 essay on the rise and decline of development economics: "we may have gained in maturity what we have lost in excitement" (33).

Recently, the World Bank reported that African women perform 90 percent of the work of hoeing and weeding (World Bank 2000). Furthermore, cassava has been described as a "women's crop" by some scholars (Ikpi 1989a, 1989b; Okorji 1983). Okorji (1983), for example, found that in the Abakaliki area of southeast Nigeria women owned more cassava fields than men, and thus concluded that cassava is a "women's crop."

COSCA researchers collected data from six countries and found that the categorization of cassava as a "women's crop" is a misleading half-truth that is based on anecdotal evidence and isolated village studies. The COSCA studies show that the proportion of the household cassava field area (hereafter field) owned by women ranged from 4 percent in the Congo to 24 percent in the Côte d'Ivoire.[7] By contrast, the proportion of cassava fields owned by men ranged from 15 percent in the Côte d'Ivoire to 72 percent in Uganda and 81 percent in Nigeria. Joint ownership by both men and women accounted for the balance of the percentages in each country. Table 2.1 points up the larger proportion of cassava fields owned by men in five of the six COSCA study countries.

There is an important exception, however, in tree crop-dominated rural economies. Among the six COSCA countries, only in the Côte d'Ivoire did women own a higher proportion of the cassava fields than did men. In the Côte d'Ivoire and in most other tree crop-dominant farming systems, the men concentrate on producing tree crops such as cocoa and coffee. The COSCA study showed that women owned significantly more cassava fields in tree crop-producing households in most of the six COSCA countries (table 2.2). Hence, although there are specific locations

**Table 2.1.** Area per household (ha) and percentage of food crops fields owned by women and men in six COSCA study countries. *Source: COSCA Study.*

| COUNTRY/CROP | HA/HOUSEHOLD | | | | PERCENTAGE | | | |
|---|---|---|---|---|---|---|---|---|
| | WOMEN | MEN | COMMON | TOTAL | WOMEN | MEN | COMMON | TOTAL |
| CONGO | | | | | | | | |
| cassava | 0.01 | 0.05 | 0.18 | 0.24 | 4 | 21 | 75 | 100 |
| yam (or sweet potato) | – | – | – | – | – | – | – | – |
| banana (or plantain) | – | – | – | – | – | – | – | – |
| maize | – | – | – | – | – | – | – | – |
| rice | – | – | – | – | – | – | – | – |
| TOTAL | 0.01 | 0.06 | 0.19 | 0.26 | 4 | 23 | 73 | 100 |
| CÔTE D'IVOIRE | | | | | | | | |
| cassava | 0.22 | 0.14 | 0.57 | 0.93 | 24 | 15 | 61 | 100 |
| yam (or sweet potato) | – | 0.05 | 0.11 | 0.17 | 6 | 29 | 65 | 100 |
| banana (or plantain) | 0.02 | 0.07 | 0.10 | 0.19 | 10 | 37 | 53 | 100 |
| maize | – | 0.05 | 0.13 | 0.18 | 0 | 28 | 72 | 100 |
| rice | 0.06 | 0.08 | 0.45 | 0.59 | 10 | 14 | 76 | 100 |
| TOTAL | 0.31 | 0.39 | 1.36 | 2.06 | 15 | 19 | 66 | 100 |
| GHANA | | | | | | | | |
| cassava | 0.11 | 0.59 | 0.45 | 1.15 | 10 | 51 | 39 | 100 |
| yam (or sweet potato) | 0.02 | 0.47 | 0.06 | 0.55 | 4 | 85 | 11 | 100 |
| banana (or plantain) | 0.04 | 0.17 | 0.04 | 0.25 | 16 | 68 | 16 | 100 |
| maize | 0.10 | 0.32 | 0.24 | 0.66 | 15 | 49 | 36 | 100 |
| rice | 0.00 | 0.06 | 0.00 | 0.06 | 0 | 100 | 0 | 100 |
| TOTAL | 0.27 | 1.62 | 0.80 | 2.69 | 10 | 60 | 30 | 100 |
| NIGERIA | | | | | | | | |
| cassava | 0.07 | 0.51 | 0.05 | 0.63 | 11 | 81 | 8 | 100 |
| yam (or sweet potato) | – | 0.19 | 0.02 | 0.21 | 0 | 90 | 10 | 100 |
| banana (or plantain) | – | – | – | – | – | – | – | – |
| maize | 0.00 | 0.21 | 0.02 | 0.23 | 0 | 91 | 9 | 100 |
| rice | 0.00 | 0.12 | – | 0.12 | 0 | 100 | 0 | 100 |
| TOTAL | 0.08 | 1.03 | 0.09 | 1.20 | 7 | 86 | 7 | 100 |
| TANZANIA | | | | | | | | |
| cassava | 0.04 | 0.52 | 0.03 | 0.59 | 7 | 88 | 5 | 100 |
| yam (or sweet potato) | 0.03 | 0.01 | 0.01 | 0.05 | 60 | 20 | 20 | 100 |
| banana (or plantain) | – | 0.03 | 0.01 | 0.04 | 0 | 75 | 25 | 100 |
| maize | 0.03 | 0.23 | 0.02 | 0.28 | 11 | 82 | 7 | 100 |
| rice | 0.01 | 0.05 | 0.01 | 0.07 | 14 | 72 | 14 | 100 |
| TOTAL | 0.12 | 0.83 | 0.08 | 1.03 | 12 | 80 | 8 | 100 |
| UGANDA | | | | | | | | |
| cassava | 0.05 | 0.25 | 0.05 | 0.35 | 14 | 72 | 14 | 100 |
| yam (or sweet potato) | 0.05 | 0.05 | 0.01 | 0.11 | 45 | 45 | 10 | 100 |
| banana (or plantain) | 0.03 | 0.11 | 0.01 | 0.15 | 20 | 73 | 7 | 100 |
| maize | 0.02 | 0.05 | – | 0.08 | 25 | 63 | 12 | 100 |
| rice | 0.00 | 0.01 | 0.00 | 0.01 | 0 | 100 | 0 | 100 |
| TOTAL | 0.15 | 0.47 | 0.08 | 0.70 | 21 | 67 | 12 | 100 |

**Table 2.2.** Cassava Field Area (ha/household) Owned by Women and Men in Tree Crop-Producing and Non-Tree Crop-Producing Households in Six COSCA Study Countries. *Source: COSCA Study.*

| COUNTRY | | WOMEN | MEN | COMMON |
|---|---|---|---|---|
| Congo[a] | tree crop-producing households | 0.00 | 0.00 | 0.07 |
| | non-tree crop-producing households | 0.01 | 0.06 | 0.20 |
| Côte d'Ivoire[b] | tree crop-producing households | 0.22 | 0.14 | 0.57 |
| | non-tree crop-producing households | – | – | – |
| Ghana[b] | tree crop-producing households | 0.08 | 0.49 | 0.43 |
| | non-tree crop producing households | 0.02 | 0.86 | 0.53 |
| Nigeria[b] | tree crop-producing households | 0.08 | 0.47 | 0.03 |
| | non-tree crop-producing households | 0.01 | 0.73 | 0.10 |
| Tanzania[b] | tree crop-producing households | 0.03 | 0.46 | 0.00 |
| | non-tree crop-producing households | 0.04 | 0.54 | 0.04 |
| Uganda[c] | tree crop-producing households | 0.09 | 0.31 | 0.08 |
| | non-tree crop-producing households | 0.01 | 0.18 | 0.03 |

[a] the tree crop is oil palm.
[b] the tree crops are cocoa, coffee, or oil palm. Every COSCA study village in Côte d'Ivoire produced at least one tree crop.
[c] the tree crops are cocoa or coffee.

in certain countries where women own more cassava fields than do men (Okorji 1983; Ezumah and Domenico 1995), these locations are exceptions and not the norm.

The COSCA study also found that cassava yields were lower (11.0 tons per hectare) in women's fields as compared to an average of 13.2 tons per hectare among men's fields for the six COSCA countries. The lower cassava yield in the women's fields is due to lower soil fertility and to earlier harvesting of cassava in these fields relative to men's fields.[8] Previous studies have shown that women are often allocated land of inferior soil fertility (Palmer-Jones 1991). The COSCA study found that, on average for the six countries, the mean age of cassava fields at harvest was lower among women's fields (12.9 months after planting) than among men's fields (15.7 months after planting). Women often grow cassava near the farm households because of easy access to cassava roots for food preparation and convenience in weeding in combination with housekeeping and childcare.

The COSCA studies show that the assertion that cassava is a "women's crop" in Africa is not valid, because men are more important in terms of the ownership of cassava fields. The myth is also rejected because in terms of work in cassava fields, men produce more cassava than women in most of the COSCA-studied households. We shall show in figure 10.2 that as a country moves through the cassava transformation stages, men devote more time to cassava production, processing, and marketing.

To summarize, the common assertion that cassava is a women's crop is a half-truth. The whole truth is that cassava is also a men's crop. As the cassava transformation proceeds, the division of labor between women and men changes and men play increasingly significant roles in cassava production, processing, and marketing. That cassava is a men's crop is an important half-truth. The same holds true for women. Together, the two half-truths point up the need to understand the dynamics of the changing division of labor between men and women as the cassava transformation proceeds.

## Cassava as a Lethal Food

Cassava contains cyanogenic glucosides in the form of linamarin and, to a lesser extent, lotaustralin. The amount of cyanogenic glucosides varies with the part of the plant, its age, variety, and such environmental conditions as soil moisture, temperature, and the like (Nartey 1977). Some varieties have been long designated as sweet or bitter,[9] purportedly in relation to their cyanogenic glucoside content. The sweet varieties are believed to be lower in cyanogenic glucosides than are the bitter varieties. However, chemical analysis of various parts of the cassava plant at different stages of development indicates that, at times, no significant differences exist between comparable parts of sweet and bitter varieties. The phellodam of sweet varieties may contain cyanogenic glucoside, while the fleshy cotex may contain none (Nartey 1977). However, the traditional methods of cassava processing and preparation as food usually remove most of the free cyanide, thus eliminating the danger of toxicity. Cases of cyanide exposure (Osuntokun 1981) and acute intoxication have been reported, but such cases are extremely rare. Potentially more serious is the possibility of chronic toxicity associated with the habitual consumption of large quantities of cassava products or with consumption of insufficiently processed cassava when the diet is deficient in protein.

Osuntokun (1973) provided evidence that indicated chronic cyanide intoxication of dietary origin as the major etiological factor in tropical ataxic neuropathy (TAN) in Nigerians.[10] Yet most Nigerians who consume cassava do not develop the disease, because of the way the cassava is prepared,[11] because of variation in biological susceptibility among individuals, and because of variations in consumption of other food items and supplements, especially animal protein.

Cyanide exposure from insufficiently processed cassava has been implicated as a factor in paralysis of both legs (*epidemic spastic paraphrasis*) reported in drought-stricken areas of Mozambique where the population consumed high levels of insufficiently dried cassava roots and a low level of protein supplements (Rosling 1986). Here, however, the toxicity of the cassava can be seen to be the lesser of two evils. Cassava was the only crop to survive drought in the early 1980s. Owing to a lack of other food, the roots, which had a high cyanide content, were eaten without the normal soaking and sun-drying to reduce toxicity. As the drought progressed, a shortage of cassava also occurred and *hunger-edema* and deaths from starvation were observed in the area but no more cases of paralysis. The extensive cultivation of toxic cassava varieties in the area thus proved to be a safeguard against famine, saving thousands from dying of starvation. Although cyanide exposure from insufficiently processed cassava may have been implicated in the deaths of some in the drought-stricken area, the bottom line is that without cassava, the entire population could have been decimated by famine.

Since cassava constitutes a staple crop with a high yield and known resistance to drought and pests, a reduction of its cultivation due to fear of toxic effects may deprive families of badly needed calories. If, on the other hand, cassava productivity and farm incomes can be raised, more families will be in a position to purchase the protein-rich foods they so badly need (Omawale and Rodrigues 1980). Since caloric intake is still a limiting factor in the diet of many people in Africa, cassava can play a very important role in meeting family food security needs.

Only under very special circumstances are toxic effects to be feared. It is important to publicize these effects in order to minimize their occurrence, yet the fact of their existence should not discourage low-income smallholders from taking full advantage of the agricultural potential that cassava offers to them (Rosling 1986).

To summarize, some cassava varieties contain cyanogens. Yet it is scientifically unjustified to condemn cassava as a lethal food, unfit for human consumption. With proper processing, the cyanogen level can be reduced to a tolerable range and cassava can become a "normal" part of a family's consumption pattern.

## Cassava as a Nutritionally Deficient Food

There is a long-standing debate on the nutritional value of cassava because it contains only 1 or 2 percent protein and is low in minerals and some essential vitamins. Some nutritionists have argued that since the diets of most Africans are low in protein, the consumption of cassava should be discouraged (Brock 1955; Nicol 1954). Latham (1979) argues that cassava's increasing popularity and spread should be a cause for concern because it contains primarily carbohydrates. Of these carbohydrates, 64 to 72 percent are made up of starch (table 2.3). About 17 percent sucrose is found in sweet varieties, and small quantities of fructose and dextrose have been reported (Hendershott et al. 1972).

Peeling results in the loss of part of the valuable protein content of the root because the peel contains more protein than is found in the root flesh. Fermentation of the roots results in protein enrichment to 6 percent (S. K. Hahn 1989). Cassava root is reasonably rich in calcium and vitamin C, but large proportions of thiamin, riboflavin, and niacin are lost during processing.

Cassava leaves are much richer in protein than the roots. Although the leaves contain far less methionine than the roots, the levels of all other

**Table 2.3. Nutrient Composition of Fresh Cassava Roots and Cassava Leaves (per 100 grams of edible portion).** *Source: FAO Food Composition Tables.*

| NUTRIENT | UNIT | CASSAVA ROOTS | CASSAVA LEAVES |
|---|---|---|---|
| Food energy | cals | 146 | 62 |
| Water | gms | 62.5 | 80.5 |
| Carbohydrate | gms | 34.7 | 9.6 |
| Protein | gms | 1.2 | 6.8 |
| Fat | gms | 0.3 | 1.3 |
| Calcium | mgs | 33 | 206 |
| Iron | mgs | 0.7 | 2.0 |
| Vitamin A | I.U. | tr | 10,000 |
| Thiamine, B1 | mgs | 0.06 | 0.16 |
| Riboflavin, B2 | mgs | 0.03 | 0.30 |
| Niacin | mgs | 0.06 | 1.80 |
| Vitamin C | mgs | 36 | 265 |

essential amino acids exceed the FAO's recommended reference protein intake (Okigbo 1980).

However, the low level of protein in cassava need not be a cause for concern if cassava is eaten with supplementary foods. The protein problem is dramatized when consumers are so poor that they are unable to afford to purchase other foods in order to achieve a balanced diet. For example, in the Congo, cassava root consumption contributed an average of 1,043 calories per day per person or about 55 percent of total daily calorie intake in 1992–96 (FAOSTAT). Cassava is the food security crop for most people in the Congo, and many people probably would not survive without it.

Which is more important: calories or protein? Platt, a nutritionist, has argued that "the first stage in the arithmetic is not in fact protein arithmetic but calorie arithmetic. A first step is to secure more food for the people so they can do more work" (Platt 1954, 136). A COSCA case study in the Congo illustrates this point.

In 1991, the lack of a balanced diet and cassava toxicity were on graphic and unforgettable display in a COSCA village in the Businga area of the Equator Province in the Congo. The village was only twenty-three kilometers from the Businga urban center, but with a four-wheel drive vehicle it took the COSCA study team three hours to travel that distance to the village. The COSCA researchers found that virtually every woman had goiter, a symptom of shortage of protein and iodine in a cassava-dominated diet. Every man suffered from hernia from the grueling job of clearing the forest for cassava cultivation. All of the children in the village had swollen toes, owing to infestation with the chigger (jigger) parasite; most were able to walk only with the aid of a walking stick. Skin problems were rampant because the village people drew their domestic water from the Congo River. Seeing the children fight over an empty can of tinned fish was a very emotional experience.

The quality of life was pathetically low in the village, yet the villagers surrendered the best huts to COSCA researchers for the night. The COSCA researchers chose not to sleep, however, because of swarms of mosquitoes. In spite of the tropical heat and humidity, the researchers made a big fire in the open and sat up all night to ward off the mosquitoes. The children played in the moonlight late into the night, as if they were immune to the pain of their chigger-swollen toes. In the morning, the village people sent one of the farmers across the river by canoe to buy chicken and plantain, which they cooked for the COSCA researchers.

How does one assess the role of cassava in this Congolese village? One could praise cassava for providing a cheap source of calories to people living in a Stone Age environment or one could criticize it because it is deficient in protein. After interviewing rural families in 281 villages across Africa, COSCA researchers pragmatically praise cassava for its contribution as the cheapest source of calories of any food grown in Africa. Yet farmers and traders in this Congolese village need improved year-round roads so that they can earn cash by selling cassava to urban consumers and use some of the increased income to purchase protein-rich foods.

The conclusion that follows from these observations is that it is a mistake to classify cassava as a nutritionally deficient food. Instead of being criticized, cassava should be praised because it is the cheapest source of calories for the poor. Moreover, cassava can be supplemented with other foods rich in protein. We agree with nutritionists that cassava is deficient in protein and vitamins, but instead of discouraging cassava production and consumption because the plant is low in protein, policy makers should focus on increasing the incomes of the poor so that they can purchase protein-rich foods to supplement their diets.

## Conclusions

We have examined five strongly held cassava myths and half-truths. The myth that cassava is primarily a subsistence crop was valid in the past, when 90 to 95 percent of the people of Africa were occupied in farming. Currently, however, 40 percent of the African people live in urban areas, while only 60 percent live in rural areas. Africa is changing and cassava is becoming an important cash crop. Today, cassava plays five important roles in African development: famine-reserve crop, rural food staple, cash crop for both rural and urban households, and, to a minor extent, raw material for feed and chemical industries. The COSCA data show that one-third to two-thirds of cassava planted in Africa is destined for urban markets. In Ghana, for example, roughly 60 percent of the cassava planted is being sold as a cash crop. The COSCA study shows that cassava is a food staple in low-, middle-, and upper-income households in both rural and urban areas.

We have marshaled farm-level and experimental evidence to discredit the myth that cassava is a soil degrader. The COSCA soil studies show that soils in cassava fields, some of which have been under continuous cultivation for at least ten years, are as fertile as soils of fields under cultivation with other crops.

The strongly held view by many donor agencies and NGO representatives that cassava is a "women's crop" is an important half-truth. Equally important is the half-truth that cassava is also a "men's crop." The COSCA studies have shown that both men and women produce cassava. Furthermore, in most countries women do not produce more cassava than do men, in terms of cassava field ownership or work in cassava fields. Men are increasingly involved in cassava production, processing, and marketing as the cassava transformation process unfolds in Africa.

The concern that some cassava varieties contain cyanogens, which are lethal, is also a half-truth. The cyanogens can be eliminated during processing by using well-known traditional processing methods such as peeling, grating, and toasting or soaking in water for four or five days and drying in the sun. Several other crops, including yam, taro, banana (or plantain), beans, and peas are also lethal if eaten without proper preparation. Today, the cases of cyanide poisoning are rare, and the fear of it should not discourage public or private investment in the cassava food economy.

Many critics claim that cassava is a nutritionally deficient food because of its low protein and vitamin content. However, one does not declare beef to be a nutritionally deficient food because it has a low carbohydrate content. The level of carbohydrate in cassava is an advantage in Africa because it makes cassava the cheapest source of food calories. Without question, the challenge ahead is to increase the efficiency of cassava production, harvesting, and processing in order to drive down the cost of cassava to consumers, especially the poor. This is an important but neglected issue in food policy debates in Africa. We reject the myth that cassava is a nutritionally inferior food.

These five myths and half-truths constitute a great deal of misinformation which discourages African governments and donors from investing in speeding up the cassava transformation. It is hoped that the COSCA findings will help African policy makers and scientists to recognize the important role cassava can play in anti-poverty and food security programs.

CHAPTER 3

# Cassava's Multiple Roles

## Introduction

Cassava plays a number of different but equally important roles in African development, depending on the stage of the cassava transformation in a particular country. Among these roles are: famine reserve, rural food staple, cash crop, urban food staple, industrial raw material, and livestock feed. However, the bulk of cassava production in Africa is consumed as food. After accounting for waste, 93 percent of Africa's cassava production in the mid-1990s was consumed as food, 6 percent was used as livestock feed, and only 1 percent was used as industrial raw material. By contrast, 48 percent of the cassava production in Asia during the same period was consumed as food, 40 percent used for export, 8 percent used as livestock feed, and 4 percent used as industrial raw material (International Fund for Agricultural Development and Food and Agriculture Organization of the United Nations 2000). In this chapter, we shall discuss the three major roles that cassava is currently playing in African development: famine-reserve crop, rural food staple, and cash crop for urban consumers. We shall show that because of inefficient production and processing methods, only a small percentage of Africa's total cassava production is used as livestock feed or as an industrial raw material.

In Africa, most of the scientific research and donor attention has focused on cassava's role as a food crop. Yet we shall show that the heavy emphasis on cassava as a food crop is selling cassava short. Cassava has an untapped potential as a livestock feed and a source of raw material (e.g., cassava starch) for industry. The challenge ahead is to intensify public and private sector research on new uses for cassava, which can enhance cassava's contribution to development by creating jobs in rural areas as well as earning foreign exchange.

### Famine-Reserve Crop

Cassava's role as a famine-reserve crop is one of the main reasons why the crop is so popular in Africa. Cassava is ideally suited to this role. It can be planted at any time of the year, and its harvesting can be delayed, if necessary, without a major change in the composition and quality of roots. Since cassava roots can be left in the ground for up to four years before harvesting, cassava is used by families as a reserve in case of drought and famine. This wide flexibility in planting and harvesting enables farmers to allocate their spare time to cassava after attending to more season-bound crops. In some countries, such as the Congo and Tanzania, women often remove a few roots from a plant in the home garden for a meal and allow the plant to continue growing.

Cassava is an important famine-reserve crop in countries such as Kenya, Malawi, Tanzania, and Zambia, where maize is the primary food staple and cassava is grown as a cushion against the instability in maize production created by erratic rainfall. Figure 3.1 shows that over the period from 1966 to 1996, the annual per capita maize production in Africa fluctuated, while cassava production remained almost stable from year to year.

Cassava's year-round harvest availability, in contrast to grain such as maize, which has a harvest season of thirty to sixty days, is another important aspect of its role as a food reserve crop. Grain crops such as maize

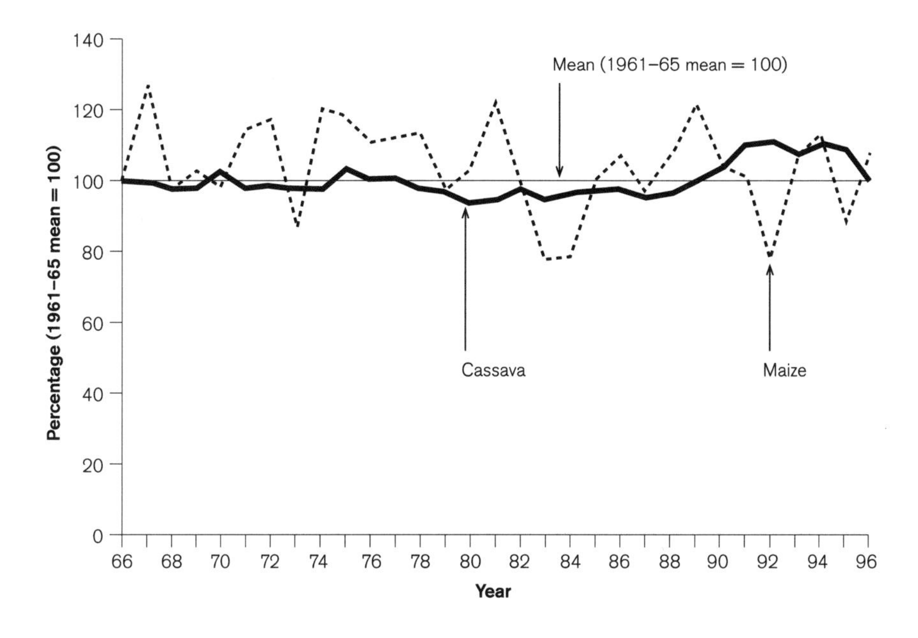


**Figure 3.1. Africa: Indices of Per Capita Production of Cassava and Maize, 1966–96 (1961–65 mean = 100 percent).** *Source: FAOSTAT.*

and sorghum are often abundant during harvesting season and scarce thereafter, leading to what is widely known as the "hungry season," a period of a few weeks or months before the harvest begins. Also, the price of crops such as maize are generally low immediately following the harvesting season and high during the hungry season, further restricting their availability to the more impoverished. In Ghana, the average monthly retail price of maize from 1991 to 1995 was significantly higher from April to July (the hungry season) than in the balance of the year. By contrast, the retail price of *gari* was stable throughout the year (fig. 3.2).

Without question, civil war and repressive governance both contribute to food insecurity. During periods of civil war, cassava has several advantages over most staple crops. The establishment cost of cassava production for home consumption is generally low because stem cuttings and family labor are the main inputs. Since the roots can be stored in the

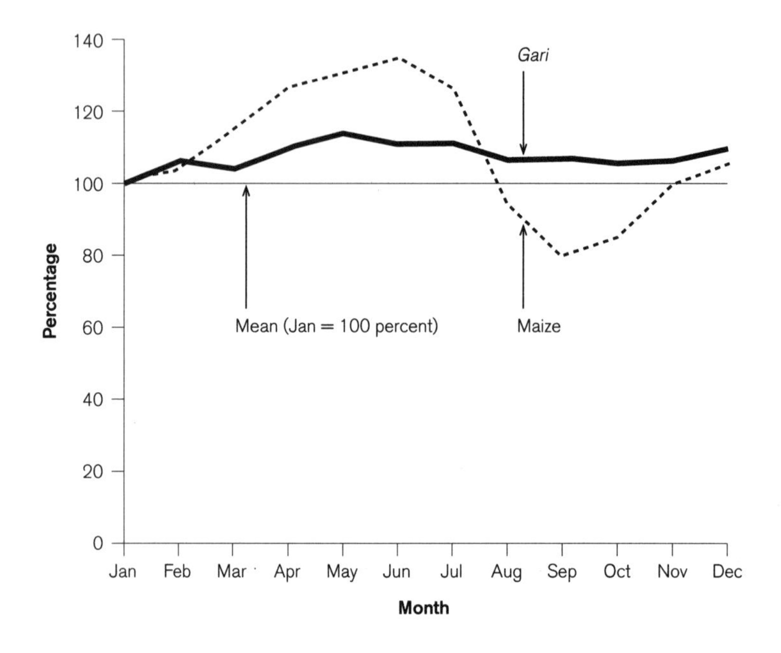


**Figure 3.2.** Ghana: Average Monthly (for 1991–95) National Retail Price Indices (January = 100 percent) for *Gari* and Maize. *Source: Policy Planning, Monitoring, and Evaluation, Accra.*

ground for several months and even up to four years without deterioration, there is a possibility that refugees can find their cassava fields unharvested upon their return home. Angola and Mozambique, for example, increased per capita cassava production from 1961 to 1997 while both were engaged in civil wars. These examples illustrate why cassava is a food reserve crop *par excellence!*

Nevertheless, despite cassava's proven role in increasing family food security and reducing poverty, most African governments are giving priority to research and extension programs on glamour crops such as maize, rice, and wheat rather than cassava.[1] Several examples illustrate this point. First, in Nigeria, government funding for research on root crops (cassava is the most important of the root crops) was three to four times more unstable than funding for research on grain from 1983 to 1995 (Idachaba 1998, 55). Second, when there is a threat of drought in countries where maize is the main food staple, such as Tanzania, Kenya, Malawi, or Zambia, the Ministry of Agriculture, often with donor support, quickly mobilizes its extension service and the NGO community to mount crash cassava production programs, which are usually abandoned soon after the drought is over. This is precisely the strategy pursued by the governments of Burundi, Ethiopia, Malawi, Rwanda, Tanzania, Zambia, and other Southern and Central African countries during the drought of the mid-1980s and the *El Nino* of the1990s (Doku 1989; International Institute of Tropical Agriculture 1995 and 1996; Mkumbira 1998; and Southern African Development Coordination Conference 1984).

## Rural Food Staple

Without question, cassava is the dominant food staple in the Congo. Cassava accounted for roughly one-half (1,043 calories) of the calories consumed per person per day over the period from 1992 to 1996 (FAOSTAT). By contrast, the second most important food crop, maize, provided only 193 calories per person per day. Furthermore, the total number of calories from cassava in local diets is likely even higher than that listed above because the FAO consumption data do not include the contribution from cassava leaves, the most important vegetable in the country. Although a high cassava-root diet is often criticized because of its low protein content (1 to 2 percent), one should keep in mind that rural poverty constrains the ability of poor consumers to purchase the protein-rich foods that are needed to achieve a balanced diet. As a result, with roughly half their daily calorie intake coming from cassava, rural people in the Congo are exposed to cassava toxicity problems.

Despite its shortcomings, cassava represents a lifeline in the Congo because it is the cheapest source of calories and it helps the rural poor to survive.[2] Yet cassava alone cannot raise the standard of living. The quality of life of the rural people of the Congo is extremely low because of decades of corruption, incompetent political leadership, and rudimentary rural infrastructure. The COSCA team's experience is illustrative of the desperate poverty and the lack of year-round roads in the Congo. During the field survey in the Bandundu Province, the COSCA study team met a sick village chief who was being head-carried by his subjects to a clinic fifty kilometers away. This mode of transport was necessary because of the lack of trucks and "bush taxis" (old station wagons). The COSCA study team was in a four-wheel-drive vehicle moving along a grass footpath at about fifteen kilometers per hour, and the carriers of the sick chief requested a ride for the chief.

Cassava is typically the dominant food staple among rural farmers who grow labor-intensive tree crops such as cocoa in West Africa and coffee in Central and East Africa. Farmers who grow tree crops as cash crops often grow cassava as a food crop for home consumption because it generates a high yield of carbohydrates per hectare and it requires labor only at planting and harvesting. For similar reasons, commercial yam farmers in Nigeria's yam belt also grow cassava as their food staple.

Yet cassava's role as a rural food staple in tree crop and commercial yam production areas can be expanded to include a cash crop role as urban demand for cassava products increases. For example, from 1940 to 1950, the growth of oil-palm production in the Ngwa area of southeast Nigeria absorbed male labor, leaving food production increasingly to women, who planted cassava because of its high yield. The growth of towns and rural incomes throughout the region created a growing urban demand for food, which, in turn, attracted young men into cassava processing (Martin 1984). In the course of the COSCA field studies in Nigeria, the study team encountered a commercial yam-producing village in the Abakaliki area in the southeastern part of the country where the farmers had all agreed to intercrop cassava and yam because the income they earned from intercropping was higher than the income they had previously earned from monocropping of yam.

### Cash Crop and Urban Food Staple

The increasing urbanization and consumer demand for convenient food products in Africa have stimulated the urban demand for cassava products such as *gari* and *attieke*. To meet the expansion in demand, cassava

**Table 3.1. Percentage Distribution of Cash Income of Cassava-Producing Households by Source in the Côte d'Ivoire, Ghana, Nigeria, Tanzania, and Uganda, 1992.** ***Source: COSCA Study.***

| SOURCE | CÔTE D'IVOIRE | GHANA | NIGERIA | TANZANIA | UGANDA |
|---|---|---|---|---|---|
| Cash Income Per Capita (US$) | 114 | 108 | 177 | 51 | 48 |
| | | | PERCENTAGE | | |
| Food crops | 15 | 55 | 55 | 16 | 28 |
| Cassava | 5 | 13 | 12 | 4 | 6 |
| Maize | 2 | 13 | 8 | 3 | 8 |
| Rice | 5 | 3 | 6 | 2 | 2 |
| Yam | 1 | 5 | 8 | 0 | 0 |
| Cooking banana | 0 | 0 | 0 | 3 | 2 |
| Plantain | 2 | 4 | 1 | 0 | 1 |
| Sweet potato | 0 | 0 | 1 | 0 | 2 |
| Other food crops | 0 | 17 | 19 | 4 | 7 |
| Other crops[a] | 76 | 21 | 20 | 25 | 11 |
| Livestock | 3 | 3 | 7 | 4 | 7 |
| Non-farm | 6 | 21 | 18 | 55 | 54 |
| TOTAL | 100 | 100 | 100 | 100 | 100 |

Note: US$1.00 = 266CFA; 430Cedi; 17Naira; 285Tsh.; 1118Ush.
[a]Other crops means industrial crops such as cotton and tree crops such as cocoa, coffee, and oil palm.

has emerged in many countries as an important cash crop for farmers and as an urban food staple. The COSCA team found that in Nigeria and Ghana, cassava serves three roles: famine-reserve crop; rural food staple; and cash crop, produced for sale to urban consumers. Table 3.1 shows that cassava is an important source of farm income in the COSCA study countries. In both Nigeria and Ghana, cassava cash incomes were highest among farms with access to mechanized cassava processing equipment for the preparation of *gari* (table 3.2).

The contribution of cassava to farm incomes was found to be low in Tanzania and Uganda because the majority of farmers in these countries lacked access to mechanized cassava processing equipment. While the majority of farmers in Tanzania prepare cassava as dried roots, most farmers in Uganda consume the roots in fresh form. Since cassava roots are bulky (70 percent water content) and perishable, they are costly to transport by truck to urban markets. Therefore, in the Congo, cassava is

**Table 3.2.** Ghana and Nigeria: Average Cassava Cash Income per Farm Household Member by Type of Processing and Preparation, 1992. *Source: COSCA Study.*

| METHOD | | GHANA (US$) | NIGERIA (US$) |
|---|---|---|---|
| PROCESSING | mechanized | 138 | 245 |
| | manual | 112 | 174 |
| PREPARATION | *gari* | 145 | 247 |
| | dried roots | 107 | 128 |

Note: US$1.00 = 430 Cedi; US$1.00 = 17 Naira.

usually processed and marketed as a dried root, which is loaded in boats along the Congo River and transported to the capital city of Kinshasa. Dried cassava roots have a limited market, however, because they require elaborate cooking—pounding, boiling, and stirring. As a result, dried cassava roots are not as attractive as *gari* to urban consumers.

This discussion illustrates the need for measures to increase the productivity of the cassava food system in order to raise the income of farmers and those involved in processing and marketing and reduce the real cost of cassava for consumers over time.

### Industrial Raw Material

Cassava starch is a raw material used in numerous industries, but it faces stiff competition from starch that is derived from numerous other plant sources, including maize, potato, sago, wheat, sorghum, rice, and arrowroot. Starch from many different sources can in most cases be combined, and it can be chemically modified to meet specific needs (Phillips 1973).

Cassava starch is an important source of industrial raw material in Asia and Latin America (International Fund for Agricultural Development and Food and Agriculture Organization of the United Nations 2000). However, it does not play a significant role as a source of industrial raw material in Africa because industries that use starch as raw material are few, and those that do exist are small. Hence, the demand for starch as a raw material is low in most African countries. Added to that is the problem of the high cost of cassava, which results from inefficient cassava production and processing methods.

In Nigeria, industries using starch include petroleum drilling, iron mining, textile, adhesives, packaging, food manufacture, pharmaceuticals, and batteries. These industries had a combined annual demand in 1989 for 60,000 tons of starch, which was equivalent to 300,000 tons of fresh cassava roots and represented only 1.5 percent of Nigeria's annual cassava production (Nweke 1992). In Ghana, 0.2 percent of the total cassava production in 1988 would have met the entire demand for starch by all the textile mills in the country, even if the mills had utilized only local cassava starch (Al-Hassan 1992).

In 1989, the five major cassava starch plants in Nigeria had a combined capacity to utilize 375,000 tons of fresh cassava roots and produce 75,000 tons of starch per year.[3] Yet the largest plant was operated at only 20 percent of its installed capacity because of an inadequate and irregular supply of cassava roots. A number of cassava starch factories have been set up as government factories in countries such as Madagascar, Tanzania, and Uganda over the last two decades, yet these, too, are no longer operational because of an inadequate supply of cassava roots (International Fund for Agricultural Development and Food and Agriculture Organization of the United Nations 2000). The bottom line is that locally produced cassava starch is not competitive with imported starch made from other plant sources because of high cassava production, processing, and transportation costs within Africa.

Cassava is an ideal crop for the manufacture of biologically degradable polymers (bioplastics) made from renewable primary materials (Stoeckli 1998). Yet the use of cassava for this purpose will require new cassava varieties and new processing technologies.

### Livestock Feed

Historically and currently, cassava plays a minor role as an ingredient in livestock feed in Africa because cassava is often more expensive than imported maize for this purpose. In Africa, only 6 percent of total cassava production was used as livestock feed in the mid-1990s. By contrast, in Latin America about half of the total cassava production during the same period was utilized as livestock feed (International Fund for Agricultural Development and Food and Agriculture Organization of the United Nations 2000).

Nigeria provides an interesting case study of how government policies can influence the use of cassava as a livestock feed. In 1985, the Nigerian government banned the importation of maize and compelled livestock feed mills to look for local crop sources such as cassava. As a result, the proportion of total cassava production used as livestock feed increased to 10 percent from 1991 to 1995, up from 3 percent in the period from 1986 to 1990 (FAOSTAT). Although the ban on maize importation was later lifted, it lasted long enough for the feed mills to make adjustments in their mills to utilize cassava pellets. Other COSCA study countries that use cassava in livestock feed are Côte d'Ivoire, Ghana, and Uganda.

The share of African cassava production used as livestock feed is probably underestimated because cassava roots and leaves are fed to pigs on small-scale farms in the cassava-producing areas, either fresh or in cut-and-dried form (International Fund for Agricultural Development and Food and Agriculture Organization of the United Nations 2000). COSCA researchers found that there is a complementary relationship between cassava production and sheep and goat rearing because cassava processing is carried out around homes, and sheep, goats, and chickens are fed by-products of cassava processing.[4]

Africa's cassava pellets are presently uncompetitive in European livestock feed markets because of the high cost of production and transportation in Africa and from Africa to Europe and because Africa has been an unreliable supplier of pellets (Phillips 1973). Cassava use in livestock feed in Europe rose from 1.5 million tons (dry weight equivalent) in the early 1970s to 7.0 million tons in 1989, before declining to 3.6 million tons in 1994. The high level of use in the 1980s was due to high domestic grain prices in the European community, which made it possible for cassava pellets from Asia and Latin America to compete effectively with domestic grain in the European market. However declining grain prices in Europe in the 1992 reform of Common Agricultural Policies (CAP) provided incentives for a greater utilization of grain by the feed industry, thereby depressing the demand for imported cassava pellets, especially from Thailand (International Fund for Agricultural Development and Food and Agriculture Organization of the United Nations 2000).

The production of cassava in Africa for export is notoriously unstable because of weather-induced fluctuations in food production (FAOSTAT; Nweke 1992; and Rosling 1986). Africa's unstable supply of cassava discourages European buyers, who have a long history of relying on more stable Asian and Latin American suppliers. The solution is to increase the productivity of Africa's entire cassava industry and improve rural roads. Yet African cassava products are also low in quality because of the inefficient traditional processing methods, and this will also need to be addressed if these products are to become competitive in the global market.

## Conclusions

Cassava plays five important roles in African development: famine-reserve crop, rural food staple, cash crop for urban consumption, livestock feed, and industrial raw material. The first three roles currently account for 93 percent of Africa's cassava production, while the last two account for only 7 percent of this production. Africa's token use of cassava in its industries and as a foreign exchange earner in European livestock feed markets is basically one of economics.

Historically, cassava was a self-spreading innovation because it was spread from farmer to farmer and used as a famine-reserve crop. Today, cassava still performs this famine-reserve function in countries such as Tanzania where rainfall is too unreliable to assure a stable maize supply. Cassava plays a role as a central rural food staple in the forest zone in countries such as the Congo where farmers are unable to deliver cassava products to urban markets because of poor roads. Cassava also is an important rural food staple in major tree crop-producing countries such as the Côte d'Ivoire and Uganda. Tree crop farmers produce cassava for home consumption because cassava production does not have the time-bound labor requirements that tree crop production does. In fact, cassava can be planted and harvested at almost any time of the year.

In countries such as Nigeria and Ghana, cassava is playing an increasingly important role as a cash crop destined for urban markets. With access to improved processing methods, farmers are able to prepare an array of cassava food products that are suitable for consumption in the urban environment.

African cassava pellets are not currently competitive with Asian or Latin American pellets in the livestock feed industry in Europe. Furthermore, African cassava starch is not competitive with imported starch. High cost, irregular supply, and low quality stemming from inefficient traditional production and processing methods limit the ability of African cassava to compete with cassava from Asia and Latin America in global markets.

Cassava will continue to play a famine-reserve role in drought-prone rural areas because it is better suited to play such a role than any other food staple. The challenge ahead is to transform cassava into an urban food staple that can both generate rural employment and raise farm incomes. Research and development are needed to increase cassava productivity in order that it can play an expanded role as a livestock feed and industrial raw material.

To summarize, many African governments and donors are selling cassava short by focusing on using it primarily as a famine-reserve crop and a rural food staple. Programs such as the research program on cassava in Nigeria, which is focused on plant breeding, are doing little to improve the productivity of the cassava food system in order to enable it to perform a greater role in African development.

# Production Overview

## Introduction

In the early 1960s, Brazil was the world's leading cassava producer, while Africa accounted for 40 percent of world production. However, by the early 1990s, Africa produced half of the total world cassava output and Nigeria had replaced Brazil as the leader in cassava production (FAO-STAT). Two forces explain this dramatic growth in cassava. First, demand for cassava has expanded because of rapid population growth, while poverty has deepened, encouraging consumers to search for cheaper sources of calories in Africa. Second, the supply of cassava has expanded because genetic research and better agronomic practices have boosted cassava yields, especially in Nigeria and Ghana. In this chapter we shall present an overview of cassava production, covering varieties grown, area cultivated, pests and diseases, yield, and production trends.

## Varieties Grown

The vegetative characteristics that are used in identification of cassava varieties include the following: type of branching, pubescence status, leaf shape, and the color of young shoot, petiole, and root flesh.[1] In the six COSCA study countries, the most prevalent varieties grown were branching, nonpubescent, purple young shoot, and creamy inner root skin (table 4.1). Varieties with broad leaves and red petioles are the most common, irrespective of agro-ecological zone.

The COSCA researchers identified a total of 1,200 local cassava varieties (based on names used by farmers) in the 281 villages studied in the six countries. The COSCA team condensed the 1,200 local varieties into 89 groups of varieties with the same morphological characteristics. The classification factors used were: bitter/sweet; branching type; pubescence status; leaf shape; and colors of young shoot, petiole, and root flesh.[2] Fifty-three groups of varieties were found to be sweet while thirty-six were bitter.[3] Genetic diversity was higher for the sweet than for the bitter varieties.

The COSCA team also found that farmers in the Côte d'Ivoire, Ghana, and Uganda planted more area to the sweet varieties, while bitter varieties were more popular in the Congo, Nigeria, and Tanzania. However, within each country the distribution of the area planted to the bitter and the sweet cassava varieties was found to vary by agro-ecological zone. For example, the COSCA study found that in the Congo, Côte d'Ivoire, Ghana, Tanzania, and Uganda, sweet cassava varieties were dominant in the forest zone (70 percent sweet and 30 percent bitter), whereas bitter varieties were more popular in the savanna zone (30 percent sweet and 70 percent bitter). In the transition zones of the same countries, the areas planted were equally divided between sweet and bitter cassava varieties. The farmers' decision on whether to plant sweet or bitter varieties in those five COSCA countries was found to be correlated with the ease of preparing dried roots in the different agro-ecological zones. Farmers in the savanna zone processed bitter cassava as dried roots by peeling and soaking the roots and drying them in the sun. Because sun-drying is inefficient in the forest zone, however, farmers there plant sweet varieties, which they can eat without soaking, drying, or fear of cyanide poisoning.

In Nigeria, however, the agro-ecological breakdown of the sweet and bitter varieties is reversed. The areas planted to cassava in the Nigerian forest zone were 25 percent sweet and 75 percent bitter; in the transition

**Table 4.1.** **Percentage of Cassava Varieties Grown in the Côte d'Ivoire, Ghana, Nigeria, Tanzania, and Uganda by Physical Characteristics and Agro-Ecological Zone.** ***Source: COSCA Study.***

| CHARACTERISTIC | FOREST | TRANSITION | SAVANNA | AVERAGE |
|---|---|---|---|---|
| | | PERCENTAGE | | |
| Sweet | 72 | 64 | 54 | 69 |
| Bitter | 28 | 36 | 46 | 31 |
| TOTAL | 100 | 100 | 100 | 100 |
| Low branching | 35 | 39 | 38 | 37 |
| High branching | 55 | 43 | 29 | 47 |
| Nonbranching | 9 | 18 | 33 | 16 |
| TOTAL | 100 | 100 | 100 | 100 |
| Pubescent | 10 | 19 | 79 | 15 |
| Nonpubescent | 90 | 81 | 21 | 85 |
| TOTAL | 100 | 100 | 100 | 100 |
| Broad leaf | 86 | 91 | 82 | 85 |
| Narrow leaf | 14 | 9 | 18 | 15 |
| TOTAL | 100 | 100 | 100 | 100 |
| Green young shoot | 51 | 43 | 40 | 45 |
| Purple young shoot | 45 | 54 | 56 | 52 |
| Other color | 4 | 3 | 4 | 3 |
| TOTAL | 100 | 100 | 100 | 100 |
| Green petiole | 31 | 39 | 35 | 35 |
| Red petiole | 63 | 51 | 63 | 58 |
| Other color | 3 | 10 | 2 | 7 |
| TOTAL | 100 | 100 | 100 | 100 |
| White inner root skin | 15 | 18 | 15 | 15 |
| Cream inner root skin | 37 | 51 | 64 | 49 |
| Pink inner skin | 44 | 28 | 20 | 33 |
| Other color | 4 | 3 | 1 | 3 |
| TOTAL | 100 | 100 | 100 | 100 |
| White root flesh | 72 | 46 | 45 | 58 |
| Cream root flesh | 25 | 53 | 52 | 39 |
| Yellow root flesh | 3 | 1 | 3 | 3 |
| TOTAL | 100 | 100 | 100 | 100 |

zone they were 55 percent sweet and 45 percent bitter; and in the savanna zone they were 65 percent sweet and 35 percent bitter. Farmers in the Nigerian forest zone planted the bitter varieties because they have the processing technology to convert the roots into *gari,* which does not require sun-drying.

The percentage of area planted to sweet varieties is higher around market centers than in remote areas. Farmers in peri-urban centers plant sweet cassava varieties and sell fresh roots as snack food in urban centers. The COSCA study found that boiled fresh roots of sweet cassava varieties are commonly sold to urban workers as a snack or a midday meal in major cities such as Kinshasa in the Congo, Kano in Nigeria, Dar es Salaam in Tanzania, and Kampala in Uganda.

The area planted to the sweet varieties is also higher in countries where cassava is a secondary food staple. Cassava is a secondary staple in the banana (or plantain) areas of Ghana, Tanzania, and Uganda, as well as in the Nigerian yam belt, and in the maize, rice, and millet/sorghum areas in the COSCA countries. In areas where cassava is of secondary importance, fresh roots of the sweet cassava varieties are often boiled and used as a supplement to pounded plantain, banana, or yam. Cassava flour is used as supplement in maize, millet, and sorghum flour (Msabaha and Rwenyagira 1989; Otim-Nape and Opio-Odongo 1989).

Cassava is frequently a family food staple in tree crop-producing areas of Africa, such as the cocoa-producing areas of West Africa and coffee-producing areas of East and Central Africa. Because there is shortage of sunlight in the forest zone for efficient drying of cassava roots, most of the tree crop farmers plant sweet cassava varieties, which can be eaten fresh.

Sweet cassava varieties are more popular than bitter varieties in drought-prone areas. Under drought conditions, processing of the sweet varieties can be carried out in a rush, shortcutting some steps, such as sun-drying or soaking in water, which are essential for the elimination of cyanogens.

Farmers interviewed by COSCA researchers reported that there are more high-yielding, pest-resistant, and in-ground storable varieties among the bitter than among the sweet cassava types. However, there are more early bulking varieties among the sweet than among the bitter types. As the cassava transformation proceeds, an increasing percentage of the total cassava production will be processed as *gari* for urban consumption. When this occurs, the debate over bitter versus sweet cassava varieties will become irrelevant. Yet in areas where there is a precarious food situation, such as periodic famine owing to wars or drought, the sweet varieties will continue to be demanded by farmers because the roots can be consumed without soaking and drying and without fear of cyanide poisoning. Therefore, the breeding of high-yielding sweet cassava varieties will need to be continued, to serve farmers in famine- and drought-prone areas.

### Pests and Disease

The cassava mosaic virus was first identified in 1891, and in 1959, on the eve of Africa's independence, Jones reported that the cassava mosaic virus was the only major disease affecting cassava (1959). However, in the early 1970s, another disease, the cassava bacterial blight, and two pests, the cassava mealybug and the cassava green mite, emerged as major problems threatening the cassava industry in Africa (Hahn, Leuschner, and Singh 1981; Yaninek 1994).

Termites are also important pests affecting cassava. In the savanna, termites make it difficult for farmers to take advantage of the cassava plant's flexible planting schedule. Cassava fields planted early or late in the rainy season in the savanna have a poor establishment record because termites feed on the planted cassava sticks. A wide range of rodent types feed on cassava roots in the forest and the transition zones and thereby cause reduction in yield.

African farmers usually do not attempt to control the cassava pests, diseases, and termites with pesticides, because of limited access to chemicals and because it is not profitable to apply pesticides on cassava. We shall show in chapter 6 that farmers use the following agronomic practices to achieve partial control of the cassava pests and diseases: fallow rotation, crop rotation, and selection of the pests and diseases-resistant local varieties. We shall now present a brief overview of the two major pests, namely the cassava mealybug and the cassava green mite, followed by a brief discussion of the two main diseases, namely the African cassava mosaic virus disease and the cassava bacterial blight.

### ■ Cassava Pests

*The Cassava Mealybug.* The mealybug is an exotic pest that was introduced into Africa from South America in the early 1970s. The mealybug was first reported in 1973 in the Congo (formally Zaire), and it rapidly spread throughout the cassava-growing areas of Africa. In some areas it destroyed so many cassava fields and local sources of the planting materials that production practically came to a halt (International Institute of Tropical Agriculture 1992). It is spread by wind and the exchange of infested planting materials. The mealybug feeds on the cassava stem, petiole, and leaf near the growing point of the cassava plant. During feeding, the mealybug injects a toxin that causes leaf curling, slowing of shoot growth, and eventual leaf withering. Yield loss in infested plants is estimated to be up to 60 percent of root and 100 percent of the leaves (Herren 1981).[4]

Starting in 1979, the mealybug was attacked through a large-scale biological control campaign by the IITA in collaboration with numerous national and international organizations. A wasp that feeds on the mealybug was identified in South America, transferred to the IITA, and reared at an IITA research station. In order to decentralize and speed up the multiplication of the wasp, IITA scientists developed a new and simpler system that was employed by most national programs in Africa (International Institute of Tropical Agriculture 1992). The wasp was first released by airplanes over cassava-growing areas in Nigeria in 1981, and later in other countries. By 1990, the wasp had been spread by airplane and by hand sprayers in twenty-four countries in the African cassava belt (Herren et al. 1987).

Without question, the biological control of the mealybug with the aid of the wasp is one of the most important scientific success stories of the past two decades in Africa. The mealybug is now held in check across most of the cassava-producing areas of Africa. The annual gain in cassava yield from the control of the mealybug is estimated at 10.2 million tons. The benefits of this biological control program will accrue for many years. For example, the wasp-use benefits were estimated to exceed three billion U.S. dollars over a twenty-five-year period beginning in 1988 (Schaab et al. 1998).

**Table 4.2.** COSCA Countries: Percentage of Villages in which Pests and Diseases Were Observed, Percentage of Plants per Field Affected by Pests and Diseases, and Symptom Severity Scores of the Pests and Diseases, 1991. *Source: COSCA Study.*

| COUNTRY | MEALYBUG | | | GREEN MITE | | | MOSAIC VIRUS | | | BACTERIAL BLIGHT | | |
|---|---|---|---|---|---|---|---|---|---|---|---|---|
| | % VILLAGES | % PLANTS/HA | SEVERITY[a] | % VILLAGES | % PLANTS/HA | SEVERITY[a] | % VILLAGES | % PLANTS/HA | SEVERITY[a] | % VILLAGES | % PLANTS/HA | SEVERITY[a] |
| Congo | 24 | 10 | 2.3 | 59 | 38 | 2.1 | 68 | 44 | 2.4 | 45 | 27 | 1.8 |
| Côte d'Ivoire | 5 | 2 | 1.2 | 7 | 3 | 1.1 | 95 | 42 | 2.0 | 24 | 5 | 1.7 |
| Ghana | 7 | 3 | 2.4 | 7 | 2 | 2.0 | 100 | 43 | 2.2 | 10 | 5 | 1.0 |
| Nigeria | 57 | 16 | 1.5 | 31 | 9 | 1.1 | 89 | 33 | 1.4 | 86 | 28 | 1.2 |
| Tanzania | 33 | 11 | 1.8 | 92 | 51 | 1.3 | 72 | 27 | 1.3 | 23 | 7 | 1.1 |
| Uganda | 5 | 2 | 1.0 | 100 | 49 | 1.7 | 64 | 30 | 2.7 | 72 | 27 | 1.6 |

[a] Severity score measured on a 1–4 scale, 4 = maximum sympton.

The mealybug is still present in Africa, and sometimes it causes damage to cassava fields even where the wasp has been well established (International Institute of Tropical Agriculture 1992). For example, in 1991, the presence of the mealybug was reported in Nigeria in 57 percent of the COSCA villages; in Tanzania in 33 percent of the villages; in the Congo in 24 percent of the villages, in Ghana in 7 percent of the villages; and in the Côte d'Ivoire and Uganda in 5 percent each of the COSCA villages. However, the percentages of plants per field infested were low in all the COSCA study countries, and they did not seriously affect cassava yields (table 4.2). Still, the persistence of the mealybug indicates the need to continue studying the impact of the biological control program.

*The cassava green mite.* The green mite was first observed in Africa in 1971 in a suburb of Kampala, the Ugandan capital. Ugandan researchers hypothesized that the green mite had attached itself to cassava cuttings that Uganda had imported from Colombia. After it became established in Uganda, the green mite spread by wind throughout Africa's cassava belt, reaching West Africa in 1979 (International Institute of Tropical Agriculture 1992). The green mite attacks cassava leaves, sucking out the fluid content of individual cells on the leaves, and the leaves become mottled and deformed. Eventually, the leaves dry out and die, although the plant usually survives. With less leaf area for photosynthesis, however, plant growth is retarded and energy from the stems and edible storage roots is consumed, resulting in drastically reduced yields (International Institute of Tropical Agriculture 1996).

Early efforts to control the green mite during the 1970s included the use of chemicals, preventive cultivation practices, and breeding to improve the cassava's resistance to the green mite. In 1983, research began at IITA on the biological control of the green mite by selecting the insects that feed on the green mite in their original environments in Colombia. This approach duplicates the model that the IITA used to gain control of the mealybug.

In 1991 the IITA scientists imported three predator mites from South America. In 1992 they multiplied them at the IITA's Biological Control Center for Africa in the Republic of Benin. The predator mites were released in farmers' fields in the Republic of Benin in 1993. In 1994 they were reported to have spread over an area totaling fifteen hundred square kilometers in the Republic of Benin, and later to eight cassava-producing countries (International Institute of Tropical Agriculture 1994).

However, the green mite is still present in Africa. In 1991, the COSCA study found that the green mite was widespread in most of the six study countries. For example, it was observed in Uganda in 100 percent of the villages; in Tanzania in 92 percent; in the Congo in 59 percent; in Nigeria in 31 percent; and in Côte d'Ivoire and Ghana in 7 percent of the study villages. The percentages of plants per field affected were also high, reaching 51 percent in Tanzania, 49 percent in Uganda, and 38 percent in the Congo.

In the Congo, the repeated harvesting of cassava leaves for consumption as a vegetable interferes with the survival of the predator mites and increases the damage to cassava plant. Since cassava leaves are widely consumed in the Congo, the biological control of the green mite using the predator mites will depend on how farmers manage the frequency of harvesting of the cassava leaves. A recent study has shown that a harvesting interval of sixty days is the optimum to maintain the population of the predator mites (Tata-Hangy 2000).

■ Cassava Diseases

*The African cassava mosaic virus disease.* The mosaic disease is transmitted by a white fly, *Bemisia tabaci,* and by the planting of cuttings derived from mosaic disease-infected plants. In a resistant cassava variety, the mosaic disease is usually confined to only a few branches. Shoots derived from cuttings obtained from symptomless branches segregate in varying proportions of incidence of the mosaic disease (Rossel, Changa, and Atiri 1994). The mosaic disease causes chlorotic blotches, distortion of the leaves, and a reduction of the leaf area. Infected plants are estimated to sustain yield losses of 30 to 40 percent (Thresh et al.1997).

The earliest effort to control the mosaic disease was the breeding of resistant varieties at the famous Amani research station in Tanzania in the 1930s. Similar breeding programs were later set up by colonial governments in Ibadan, Nigeria; Kumasi, Ghana; Njala, Sierra Leone; and Kisangani, in the Congo. As we shall show in the next chapter, this research effort has been long and productive, spanning the work of several successive breeders over more than sixty years from the 1930s to the 1990s. The IITA scientists drew on the research results of the Amani scientists' breeding efforts in the development of mosaic disease-resistant varieties that are now widely grown in Nigeria. More time and effort are required, however, to diffuse these varieties to other cassava-producing countries in order to reduce the impact of the mosaic disease on cassava production in Africa.

In 1991, COSCA researchers tested the incidence of the mosaic disease in a large percentage of villages in each of the COSCA countries. The results were as follows: Ghana, 100 percent of villages affected; Côte d'Ivoire, 95 percent; Nigeria, 89 percent; Tanzania, 72 percent; the Congo, 68 percent; and Uganda, 64 percent. The numbers of infected cassava plants per field were also high in each country, and in most of the countries the severity of the disease was high as well.[5]

Another method of controlling the mosaic disease is the sanitation approach recommended by the virologists and vector entomologists in various countries in Africa since the 1970s (Thresh et al. 1997). This approach involves the use of mosaic disease-free planting materials and the uprooting of infected plants in the field. This method has not been effective in Africa because the farmers do not have the technology to select mosaic disease-free planting materials. Furthermore, African farmers are averse to uprooting a food plant in their fields even if the plant is unhealthy. As the cassava transformation proceeds, the private sector will hopefully develop enough to provide healthy planting materials to the farmers.

The latest effort to control the mosaic disease was through the Cassava Biotechnology Network (CBN), which was established jointly by Centro Internacional de Agricultura Tropical (CIAT) and the IITA in 1988 and sponsored by the Dutch government beginning in 1992. The network involved scientists from national and international organizations in several developed and developing countries. The research activities of the CBN covered all aspects of cassava work, including processing, agronomy, breeding, and pest and disease control (Thro 1998). Unfortunately, the CBN in Africa was terminated in 1998 when the Dutch government funding was withdrawn. The plan was to have the African CBN coordinated by the IITA, but donor financing was not available (Bokanga 2000; Mba 2000). The South American CBN, however, continues to be funded by the Dutch, and it is coordinated by the CIAT.

The COSCA study found that in the Congo, where cassava leaves are important vegetable, farmers consider mosaic disease-infected leaves more palatable than uninfected leaves. Since farmers who favor the mosaic disease-infected leaves for human consumption do not consider the mosaic disease to be a problem, they will not adopt control measures or remove mosaic disease-infected plants from the fields.

Thus, although the mosaic disease problem has been known since 1891 and has been studied since the 1930s, the disease is still a major constraint on cassava production in Africa. The future control of the mosaic disease will depend on extension efforts to diffuse the IITA's resistant varieties in the cassava-producing countries and on the development of a private sector supply market for healthy cassava planting materials.

*The cassava bacterial blight.* The bacterial blight was first identified in Africa in 1972 in the Niger Delta area of Nigeria (Persley 1977). The bacterial blight spread rapidly to all the cassava-producing areas of Africa by the early 1990s.

The bacterial blight occurs mainly during the wet seasons. Symptoms include water-soaked leaf spots, blight, leaf wilt, gum exudation from infected stems, and severe defoliation. Stem die back is one of the most severe symptoms. Yield loss on infected plants of up to 60 percent has been reported in Nigeria (Terry 1979). The bacterial blight adversely affects the supply of cassava leaves in the Congo, where cassava leaves are the most important vegetable. The bacterial blight is disseminated by the use of infected planting materials.

The appearance of the bacterial blight in the Niger Delta area of Nigeria in 1972 inspired a partnership between the Shell-BP Petroleum Development Company of Nigeria and the cassava program of the IITA to breed and diffuse improved bacterial blight-resistant varieties. This partnership contributed to the rapid development and diffusion in Nigeria of improved cassava varieties that are resistant to the bacterial blight (Ohunyon and Ogio-Okirika 1979; Heys 1977).

The bacterial blight was observed in several COSCA villages in Nigeria, Uganda, and the Congo in 1991. Yet the percentage of plants infected per field and the severity of the infections were generally low. Wide diffusion of the IITA's resistant varieties and use of uninfected planting materials will help control the bacterial blight.

## Area Cultivated

In the early 1950s, cassava occupied 50 to 80 percent of all the land planted to the starchy staples in the Congo (Jones 1959). In 1991, cassava was found to be the most important crop in the staple production system in all of the seventy-one COSCA study villages in the Congo.[6] In Ghana,

**Table 4.3.** COSCA Countries: Total Area (million ha) Planted to Cassava per Year, 1961 to 1965 Compared with 1995 to 1999. *Source: FAOSTAT.*

| COUNTRY | 1961–1965 | 1995–1999 | % CHANGE |
|---|---|---|---|
| Congo | 1.37 | 2.19 | 60 |
| Côte d'Ivoire | 0.20 | 0.33 | 65 |
| Ghana | 0.15 | 0.60 | 300 |
| Nigeria | 0.83 | 2.94 | 254 |
| Tanzania | 0.59 | 0.62 | 5 |
| Uganda | 0.29 | 0.35 | 21 |

cassava is more important in the southeastern part than in the rest of the country. In Nigeria, cassava is more important in the southern parts. In Tanzania, cassava is most important along the coastal strip of the Indian Ocean.[7]

From 1961 to 1965, 5.6 million hectares were planted each year to cassava in Africa. Thirty-five years later, from 1995 to 1999, 10 million hectares were planted each year to cassava. The six countries that currently account for most of the cassava area are the Congo, the Côte d'Ivoire, Ghana, Nigeria, Tanzania, and Uganda (table 4.3). The area planted to cassava increased almost threefold in Nigeria and Ghana from 1961 to 1999 (fig. 4.1). Farmers in most of the villages in Ghana and Nigeria cited market access as the reason for expansion of area planted to cassava, while farmers in most of the COSCA villages in the Congo cited lack of access to market centers as the reason for reducing the area planted to cassava.

One critical variable in the expansion of the cassava area in Nigeria and Ghana is the availability of improved processing equipment to remove water from the roots (the roots are 70 percent water) and thereby reduce the cost of transporting them to market. Improved processing and food preparation methods reduce bulk and make it possible for cassava products to be transported at reduced costs over poor roads to distant urban market centers. One example is the steady shipment of dried roots (*cossettes*) from the Bandundu region of the Congo to the capital city, Kinshasa, by boat along the Congo River or by trucks over extremely poor road conditions.

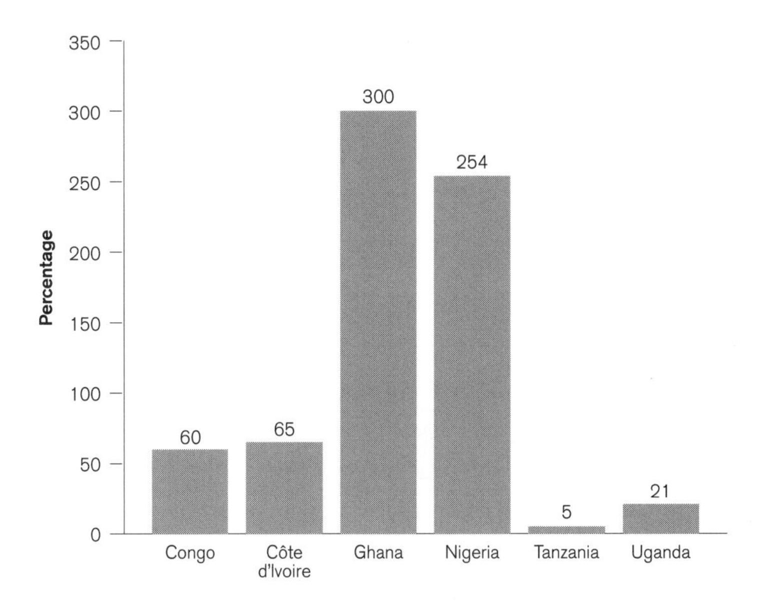


**Figure 4.1. COSCA Countries: Percentage Change in Cassava Area between 1961 and 1999.** *Source: FAOSTAT.*

Looking ahead, the expansion of cassava production will require breaking of the harvesting and processing labor bottlenecks. In Ghana and Nigeria, all the villages where farmers had access to mechanized cassava graters reported an increase in the area planted to cassava from 1970

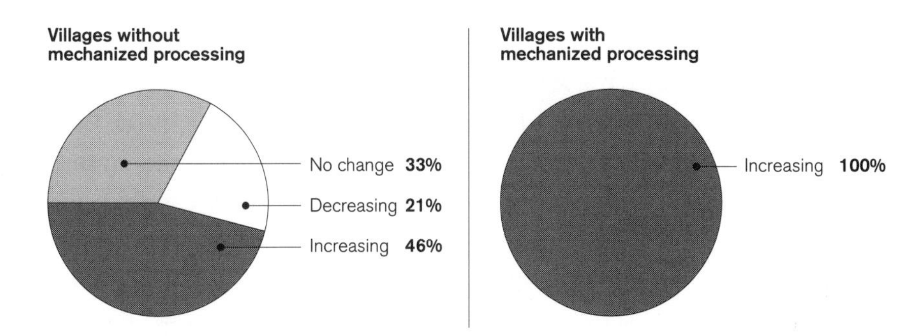


**Figure 4.2. Nigeria and Ghana: Cassava Land-Area Trends in Villages Having Mechanized Processing Compared with Villages Having No Mechanized Processing.** *Source: COSCA Study.*

**Table 4.4.** Average Cassava Yield and Yield Components in the Congo, Côte d'Ivoire, Ghana, Nigeria, Tanzania, and Uganda, 1991–92. *Source: COSCA Study.*

| YIELD COMPONENT | MEAN | MINIMUM | MAXIMUM | STD DEV. | NO. OF FIELDS |
|---|---|---|---|---|---|
| Fresh root (ton/ha) | 11.90 | 0.40 | 67.10 | 8.39 | 501 |
| Plant density (stand/ha) | 7774 | 500 | 41250 | 5280 | 500 |
| Number of roots/plant | 6 | 0 | 36 | 4 | 500 |
| Ave, root weight (kg/root) | 0.40 | 0.05 | 5.18 | 0.32 | 500 |
| Harvest index | 0.50 | 0.03 | 0.89 | 0.13 | 497 |

to 1990. By contrast, only 46 percent of the villages where farmers did not have access to a mechanized cassava grater in the two countries reported an increase in the area planted to cassava (fig. 4.2).

To summarize, the area planted to cassava has increased severalfold in two of the six most important cassava-producing countries in Africa, namely Nigeria and Ghana, between the early 1960s and late 1990s. The increase in the cassava area in Nigeria and Ghana was driven by availability of improved varieties, processing technologies, and improved road access to market centers.

### Yield

In 1954, the average cassava yield in Africa was between 5 and 10 tons per hectare (Jones 1959). The COSCA study showed that the average yield was between 10 and 15 tons per hectare in the six COSCA countries in early 1990s. Therefore, one can safely say that the cassava yield is increasing in Africa in the early 1990s because of the planting of high-yielding varieties and the adoption of better agronomic practices. In the early 1990s, the COSCA yield measurements show that the average on-farm cassava fresh-root yield (hereafter yield) for the six COSCA study countries was 11.9 tons per hectare (table 4.4).[8] The average farm-level yield was highest in Nigeria, where the mean was 14.7 tons per hectare, followed by Ghana, where the mean was 13.1 tons per hectare (fig. 4.3). The mean yield was around 10.0 tons per hectare in the Côte d'Ivoire, the Congo, Tanzania, and Uganda, respectively.

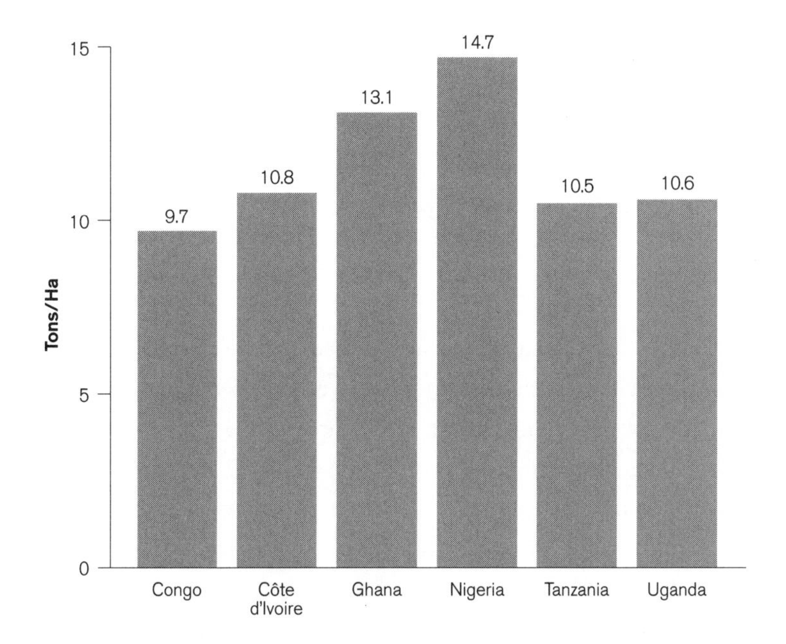


**Figure 4.3.** Cassava Yield (tons/ha) in the Congo, Côte d'Iviore, Ghana, Nigeria, Tanzania, and Uganda, 1991.

*Source: COSCA Study.*

Some experts on African farming systems have reported that the average cassava yield is stagnating or declining in Africa (Lagemann 1977; Fresco 1986). They attribute the negative trend to the increasing population pressure on land, which leads to a breakdown of the shifting cultivation system and a decline in soil fertility. The COSCA study has used two-point-in-time cassava yield data collected from three villages of low, medium, and high population densities in southeast Nigeria to challenge the assertion of declining yield owing to increasing population density.

In 1973, Lagemann studied cassava yields in three villages in southeastern Nigeria with different population densities and concluded that cassava yield declined as population pressure increased (Lagemann 1977). Lagemann found the mean yields for the three villages to be: 2.0 tons per hectare in the high population density village; 3.8 tons per hectare in the medium population density village; and 10.8 tons per hectare in the low population density village.

**Table 4.5.** Cassava Yield Trend (1973–93) in Three Southeast Nigerian Villages with Different Population Densities. *Sources: [a]Lagemann 1977; [b]COSCA Study.*

| VILLAGE POPULATION DENSITY | 1973[a] | 1993[b] | % CHANGE |
|---|---|---|---|
| High | 2.0 | 4.04 | +102 |
| Medium | 3.8 | 4.77 | +26 |
| Low | 10.8 | 9.27 | −14 |

In 1993, the COSCA study team revisited the three villages to assess the trends in the cassava yield twenty years after the Lagemann's original study. The COSCA team found that the mean yields for the villages were: 4.04 tons per hectare in the high population density village; 4.77 tons per hectare in the medium population density village; and 9.27 tons per hectare in the low population density village (table 4.5). The 1993 cross-sectional yield trend is consistent with Lagemann's 1973 observation that the yield was substantially higher in the low-population-density village than in the high-density village.

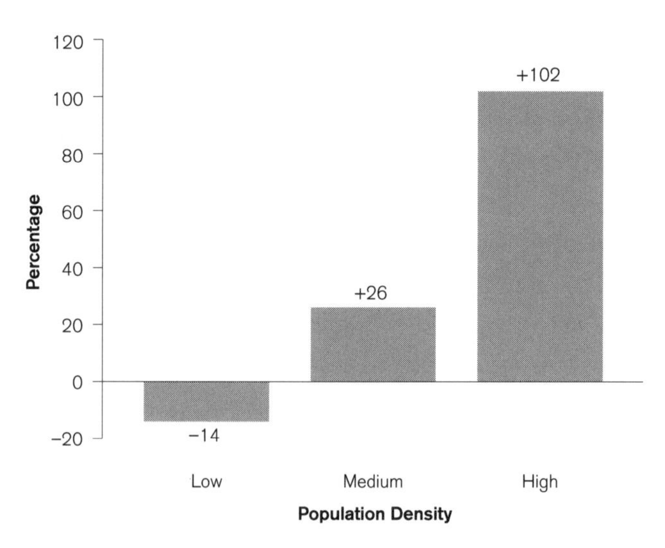


**Figure 4.4.** Southeast Nigeria: Percentage Changes in Cassava Yield from 1973 to 1993 in Low-, Medium-, and High-Population Density Villages. *Sources: COSCA Study; Lagemann 1977.*

However, between 1973 and 1993, the cassava yield doubled in the high-density village, increased by 26 percent in the medium-density village, and declined by 14 percent in the low-density village (fig. 4.4). This is a surprising result that deserves to be carefully examined. The doubling of cassava yield in the high population density village occurred because farmers planted the IITA's high-yielding cassava varieties at high stand densities, employed hired labor, and enjoyed ready access to a nearby market.[9] By contrast, yields declined in the low population density village because farmers planted local varieties at low stand densities (Enete, Nweke, and Okorji 1995). The evidence from the COSCA study demonstrates that farmers have adopted new technology such as improved varieties and improved agronomic practices as a response to increased population pressure on land. This finding adds evidence to the Boserup hypothesis that increased population pressure can induce intensification of agricultural production (Boserup 1981).

## Production

Between 1961 and 1999, total cassava production in Africa nearly tripled, from 33 million tons per year from 1961 to 1965 to 87 million tons per year from 1995 to 1999, in contrast to the more moderate increases in Asia and South America (fig. 4.5). Most of the dramatic increase in cassava production in Africa was achieved in Nigeria and Ghana. In each of these countries the production growth rate was greater than the rate of population growth. In the other COSCA study countries, the increase in cassava production kept pace with population growth.

From 1961 to 1965, Nigeria produced only 7.8 million tons of cassava per year and was the fourth-largest producer in the world after Brazil, Indonesia, and the Congo (FAOSTAT). From 1995 to 1999, Nigeria produced 31.8 million tons per year and became the largest producer worldwide, displacing Brazil, Indonesia, and the Congo. Ghana was only the seventh-largest producer in Africa from 1961 to 1965, with an annual production of only 1.2 million tons. From 1995 to 1999, however, Ghana produced 7.2 million tons annually and advanced to the position of the

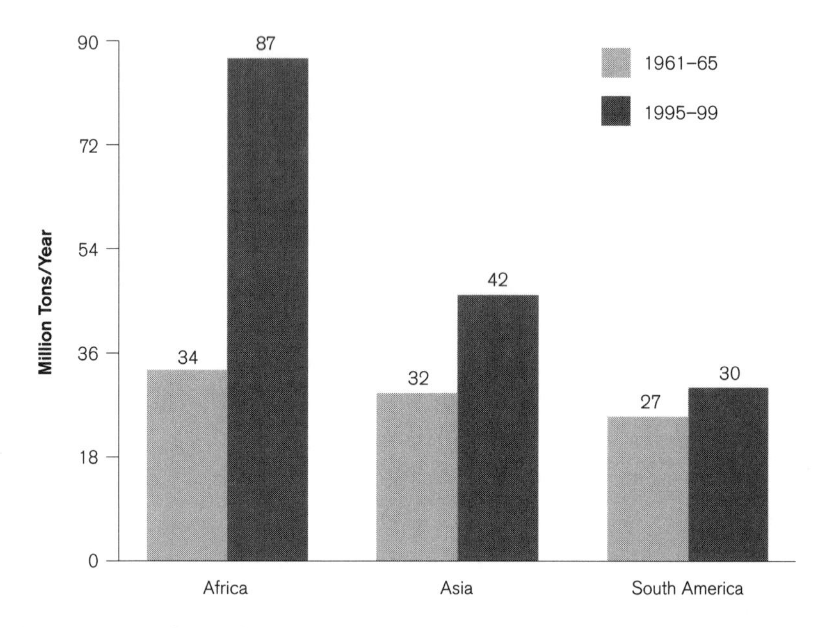


**Figure 4.5.** **Annual Cassava Production in Africa, Asia, and South America from 1961 to 1965 and 1995 to 1999.** ***Source: FAOSTAT.***

third-largest producer in Africa after Nigeria and the Congo. The dramatic increase in cassava production in Nigeria and Ghana was achieved through an increase in both area and yield. The availability of cassava graters to farmers in both countries released labor, especially female labor, from cassava processing, thus allowing them to plant more cassava. We shall show in the next two chapters that the common use of hired labor in these two countries and the wide adoption of the IITA's high-yielding varieties were responsible for higher cassava yields in Nigeria and Ghana than in the other COSCA countries.

Cassava's low input requirements, a trait that is compatible with Africa's resource endowments (relatively abundant land and seasonal labor scarcity) and the cassava's resistance to pests and diseases explain the expansion in cassava production since the 1960s. Since cassava can be grown without inputs and irrigation water, it is a low-risk crop for African farmers. Moreover, as the average farm size shrinks under population pressure, farmers are searching for crops with a higher output of energy per hectare as a strategy for overcoming hunger. Food shortages precipitated by a combination of political and civil unrest, economic stagnation, erratic rainfall patterns, and rapid population growth have had a much greater influence on cassava production in Africa than anywhere else in the world (Scott, Rosegrant, and Ringler 2000).

## Summary

Both the sweet and the bitter cassava varieties are grown in Africa. The sweet varieties are more popular in the Côte d'Ivoire, Ghana, and Uganda, while the bitter varieties are more common in the Congo, Nigeria, and Tanzania. The COSCA farmers reported that the bitter varieties are more resistant to pests, are higher yielding, and store better in the ground than the sweet varieties. Farmers in the forest zone, however, plant more sweet varieties than those in the transition and the savanna zones, because limited sunshine in the forest zone makes it difficult to sun-dry the roots after they have been soaked in order to eliminate cyanogens. Tree crop farmers also plant sweet varieties, which they eat without the soaking and sun-drying. As the productivity of the cassava system increases and more cassava is processed as *gari,* the issue of the sweet versus the bitter cassava varieties will become irrelevant.

Cassava has been under attack by pests and diseases, such as the mealybug, the green mite, the mosaic disease, and the bacterial blight. The biological control method has been used to bring the mealybug under control, while the biological control of the green mite is still in progress. The most promising method of controlling the mosaic disease and the bacterial blight is breeding resistant varieties, but this involves a long and painstaking process of breeding and diffusion. Meanwhile, the green mite, the mosaic disease, and the bacterial blight continue to cause yield losses in various parts of Africa.

Most of the major cassava pests and diseases are new in Africa, as they were introduced only within the last thirty years. Improved quarantine inspection is needed to prevent, as much as possible, introduction of new pests and diseases. There is need for better preparedness to control new problems before they spread and take root in Africa.

Between 1961 and 1999, the area planted to cassava increased many-fold in Nigeria and Ghana because of the planting of high-yielding varieties, processing technologies, and improved roads to reduce transport costs to market centers in these countries. The evidence provided by the COSCA study shows that the average cassava yield was increasing in the early 1990s in Africa because of the planting of high-yielding varieties and the adoption of improved agronomic practices in some of the COSCA countries. Therefore, although Africa's population pressure on land is increasing, the cassava yield is also increasing, particularly where cassava is planted as a cash crop for urban consumption.

The COSCA study provides new evidence on the relationship between technological change and population densities in three villages in south-east Nigeria. Over the 1973 to 1993 period, cassava yields doubled in the high population density village due to planting of improved IITA cassava varieties, which can increase yield by 40 percent without additional inputs.

CHAPTER 5

# Genetic Research and the TMS Revolution

## Introduction

African farmers and public-sector researchers have worked together in the genetic improvement of cassava which culminated in the release of the high-yielding TMS varieties in Nigeria in the late 1970s. The farmers' contribution started from the time they began to plant cassava in the sixteenth century. Farmers field-tested varieties with various attributes and exchanged these varieties among themselves and across wide areas of Africa. Government cassava research was initiated in the 1920s by the British colonial government in East and West Africa; by the French in West Africa; and by the Belgian government in Central Africa. The research agenda was an integral part of colonial campaigns to encourage farmers to plant cassava as a famine-reserve crop. In East Africa, the research financed by the British colonial government was inspired by the appearance of two cassava virus diseases: the mosaic virus and the brown streak virus. H. H. Storey assembled a research team at the Amani research station in Tanzania in the 1920s and 1930s and helped unravel the mystery of the viruses and how to control them. When the IITA established its cassava research program in 1971, Dr. S. K. Hahn drew heavily on the research on cassava mosaic virus of H. H. Storey and his colleagues (Hahn, Howland, and Terry 1980). Thus, the scientific journey that led to the TMS revolution spanned some forty years, beginning with the work of Storey in the 1930s and culminating with the research of Hahn in the 1970s.

This chapter traces the incremental nature of cassava research over the past sixty-five years (from 1935 to 2000), culminating in the release of the IITA's high-yielding TMS varieties in 1977, which were quickly adopted by farmers in Nigeria in the 1980s but continue to evolve under the leadership of A. G. O. Dixon, who succeeded Hahn at IITA in 1994. This chapter also discusses how the findings of colonial research programs in the 1930s contributed to cassava research in the national research systems (NARs) and the international agricultural research centers (IARCs). Special attention is given to the cumulative nature of cassava research and the close cooperation between "farmer-researchers" and researchers in the national research programs and international centers such as the IITA and the CIAT.

## Farmer-Led Research

Most of the cassava varieties planted in Africa today have been developed from self-sown high-yielding seedlings that are retained and planted by farmers (Doku 1966). The COSCA researchers found that farmers are quick to replace lesser varieties with those displaying superior attributes. The COSCA researchers studied the attributes desired by farmers in the six study countries and found that farmers selected varieties for the following attributes: high yield, earliness of bulking (early maturing), weed suppression ability, desirable branching habits, in-ground storability, pest and disease tolerance, low cyanogen level, ease of peeling, mealiness after cooking, drought tolerance, high leaf yield, and ease of harvesting (table 5.1).

COSCA researchers also found that the four most important attributes selected by farmers who produce cassava as a cash crop for urban markets in Nigeria and Ghana are: high yield, earliness of bulking, ease of harvesting, and ease of peeling. Farmers producing cassava as a cash

**Table 5.1.** Attributes that Farmers Used to Select Cassava Varieties in the Congo, Côte d'Ivoire, Ghana, Nigeria, Tanzania, and Uganda, 1991. *Source: COSCA Study.*

| ATTRIBUTE | CONGO | CÔTE D'IVOIRE | GHANA | NIGERIA | TANZANIA | UGANDA |
|---|---|---|---|---|---|---|
| | | | PERCENTAGE | | | |
| Earliness | 29 | 7 | 47 | 26 | 8 | 19 |
| High yield | 2 | 27 | 17 | 24 | 20 | 18 |
| Branching/weed control | 49 | 17 | 0 | 7 | 0 | 0 |
| In-ground storage | 6 | 0 | 0 | 2 | 28 | 19 |
| Pest/disease resistant | 3 | 0 | 3 | 15 | 12 | 13 |
| Harvesting and peeling | 5 | 10 | 10 | 19 | 2 | 13 |
| Low cyanogens | 0 | 27 | 10 | 2 | 7 | 3 |
| Cooking qualities | 0 | 0 | 7 | 2 | 0 | 3 |
| Others[a] | 6 | 12 | 6 | 3 | 23 | 12 |
| TOTAL | 100 | 100 | 100 | 100 | 100 | 100 |

[a]Others are yield of planting materials, leaf yield, and drought tolerance.

crop want varieties that are easy to harvest because it is difficult to recruit and manage a large itinerant labor force for manual harvesting. As a result, there is an urgent need to develop mechanized cassava harvesters.

Farmers in Nigeria and Ghana who plant cassava and convert it to *gari* for sale in urban centers select cassava varieties that are easy to peel. In these two countries, manual peeling is the most labor-intensive of the processing tasks. The roots of most of the present varieties in Africa can be peeled only by slashing the root with a sharp knife, because the skin of the root is thin. Farmers told COSCA researchers that they want varieties with a thick skin so that the peels can be easily rolled off the roots. Yet the development of mechanized cassava peeling is hampered by a lack of uniformity in the shape and size of cassava roots.

In Tanzania, farmers reported to COSCA researchers that good in-ground storability of cassava was a highly desired attribute because maize is the preferred staple and cassava is a famine-reserve crop. Farmers often leave cassava in the ground for several years, harvesting and consuming it only when they have a poor maize harvest.

In the Côte d'Ivoire, cassava is an important food staple among cocoa farmers. Cocoa farmers typically select cassava varieties with a low cyanogen content, which can be eaten without processing, because family labor can earn higher returns in cocoa farming than in processing cassava to reduce or eliminate cyanogens. In the Congo, a large canopy (profuse branching) is the most common attribute of cassava varieties selected by farmers because it is important for weed control and the leaves are harvested for human consumption.

## Government Research

### ■ Colonial Period

In 1891, Warburg reported that the mosaic (cassava mosaic virus) disease was prevalent in East Africa and adjacent islands. Soon after, the mosaic disease was reported in most countries in Central and West Africa (Storey and Nichols 1938). The widespread occurrence of the mosaic disease motivated the British colonial government to launch a cassava-breeding program at the Amani research station in Tanzania in the mid-1930s. The goal of research was to develop varieties that were tolerant to the mosaic disease.[1]

Research on varieties resistant to the mosaic disease was also carried out by the British colonial government researchers in the Coast Experiment Station in Kibarani in Kenya, the Morogoro Experiment Station in Tanzania, the Agricultural Department in Zanzibar, and the Serere Experiment Station in Uganda (Nichols 1947). Similar research was also carried out in the Kumasi research station in Ghana, at Njala in Sierra Leone, and at the Moor Plantation research station in Ibadan, Nigeria (Jones 1959).

The French colonial research on cassava was carried out by scientists at Institut De Recherches Agronomiques Tropicales (IRAT) (Fresco 1986). In 1913, the French established a research station at Bambey in Senegal for peanuts primarily for export. In 1950, the scope of research at the station was expanded to include food crops, and the research designed cassava variety selection programs with the goal of finding varieties that were high yielding and suitable for processing as *gari* (Jones 1959).

In 1933, the Belgian Government established the Institut National pour l'Etude Agronomique du Congo Belge (INEAC) at Yangambi in the

Congo to pursue research on agricultural development, including the genetic improvement of cassava in Central Africa. Nearly forty research stations were established by INEAC in Central Africa (Fresco 1986). Initially, the cassava genetic improvement objective of INEAC was to select local varieties that were best suited for small-scale processing for home consumption. After 1950, however, as the urban demand increased for cassava products such as *chickwangue* and dried root flour to serve urban consumers, the goal of research was extended to include selection of varieties suitable for intensive mechanized production (Drachoussoff, Focan, and Hecq 1993).

INEAC scientists pursued research by raising large numbers of seedlings both from controlled crosses of cassava varieties and from naturally pollinated cassava flowers and selecting the varieties with desired attributes (Nichols 1947). Genetic materials used in crosses were obtained from various external sources including Brazil and the Côte d'Ivoire, as well as from local sources. The selection criteria included reduced branching; strong stems; resistance to root rot, the mosaic virus, and lodging; short and fat roots regularly spaced around the plant; and high protein content in the root (Fresco 1986). In 1951, one of the INEAC's research stations in the Congo—Kiyaka in the Kwango-Kwilu—obtained a yield of about 45 tons per hectare with improved varieties and of 37 tons per hectare with local varieties on experimental plots without fertilizer (Fresco 1986; Drachoussoff, Focan, and Hecq 1993).[2]

Without question, the Amani research station program was the most successful colonial cassava-breeding program in Africa. In 1935, H. H. Storey conducted a worldwide search for cassava varieties that were resistant to the mosaic disease. Storey found some varieties in Java with a limited degree of resistance. Then Storey and his assistant, R. F. W. Nichols, discovered that sugar cane varieties immune to sugar cane mosaic disease were developed by crossing the sugar cane plant with its wild, non-sugar-producing relative. Hence, Storey and Nichols crossed cassava with tree species that are related to cassava genetically, namely Ceara rubber, Manicoba rubber, and "tree" cassava (Nichols 1947).[3] These species conferred resistance to the mosaic disease to their hybrids, namely

Ceara rubber × cassava, Manicoba rubber × cassava, and "tree" cassava × cassava hybrids (Jennings 1976). Yet although the various rubber species × cassava hybrids were resistant to the mosaic disease, they were not real cassava because they produced a low root yield, of poor food quality, and they had poor agronomic characteristics such as lodging.

However, during World War II (from 1939 to 1945), the breeding work at the Amani research station was scaled back (Nichols 1947). In 1951, Nichols died in an automobile accident and was replaced by D. L. Jennings.

Jennings conducted on-farm testing of the various Storey/Nichols rubber species × cassava hybrids and found that the Ceara rubber × cassava hybrids (e.g., 46106/27) provided the best combinations of yield, root quality, and resistance to both the mosaic and the brown streak diseases. Jennings also discovered that although the resistance to the mosaic disease was satisfactory for inland areas, it was frequently inadequate for coastal regions. He also found that the Manicoba rubber × cassava hybrids did not have a good yield potential (Jennings 1976).

Jennings intercrossed the Storey/Nichols mosaic- and brown streak-resistant rubber species × cassava hybrids to release recessive genes for resistance and to combine genes that had been dispersed during the process of backcrossing by Storey and Nichols. This led to segregates (e.g., 5318/34) that showed much higher and more stable resistance over a wide area than had the hybrids created by Storey and Nichols. Jennings distributed pollinated seeds of these segregates to several African countries in 1956, one year before the Amani research station program was terminated in 1957 (Jennings 1976).[4]

In 1958, at Moor Plantation research station, Ibadan, in Nigeria, B. D. A. Beck and M. J. Ekandem selected the Ceara rubber × cassava hybrid, 58308, from the seed derived from the Jennings series 5318/34. The Ceara rubber × cassava hybrid, was resistant to the mosaic disease but gave low yield and had poor root quality and poor agronomic characteristics such as lodging. Beck and Ekandem crossed the Ceara rubber × cassava hybrid, 58308, with high-yielding West African selections to combine the mosaic disease-resistance genes of the Ceara rubber ×

cassava hybrid with the high-yield genes of the West African varieties (Jennings 1976).

## ■ Postindependence Research and the TMS Breakthrough

At Nigeria's independence in 1960 the cassava breeding program at the Moor Plantation research station, Ibadan, was moved to the Federal (now National) Root Crops Research Institute, Umudike, in Eastern Nigeria and breeding work was continued by Ekandem. Unfortunately, almost all the progenies developed from the Ceara rubber × cassava hybrid (58308) and the records of the research program at Umudike, along with records that were transferred from the Moor Plantation research station in 1960, were lost during the Nigerian Civil (Biafran) War (1967–70). However the original Ceara rubber × cassava hybrid (58308) was retained at the Moor Plantation research station (Beck 1980).

As a rule of thumb it takes an average of a decade for plant breeders to develop and farmer-test a variety before it is released to extension agents. The development of the TMS cassava varieties in six years (between 1971 and 1977) thus represents an important scientific achievement and a testimony to the work of a devoted scientist, S. K. Hahn of IITA, and his colleagues.[5] Hahn's strategy for developing the TMS varieties was a collaborative undertaking involving national cassava research programs, training national scientists, developing partnerships with private companies, and investing in germ plasm exploration and conservation. The IITA's cassava-breeding program was carried out by a multidisciplinary team, including a plant pathologist, an entomologist, a nematologist, a virologist, an agronomist, a tissue culture specialist, a biochemist, and a food technologist (Dixon, Asiedu, and Hahn 1992).

Cassava breeding at the IITA commenced in 1971, when S. K. Hahn was appointed as the leader of the institute's root and tuber program. Hahn invited two of Storey's former colleagues to join his research team at IITA: A. K. Howland, who participated from 1972 to 1976 and D. L. Jennings, who participated in 1975.[6] Hahn set about developing new cassava varieties with two key characteristics: mosaic resistance and high yield. Drawing on the earlier work of Storey, Hahn and his team members combined the mosaic-resistance genes of the Ceara rubber × cassava hybrid (58308) with genes for high yield, good root quality, low cyanogen content, and resistance to lodging. Hahn utilized the Ceara rubber × cassava hybrid (58308) as a source of resistance to the mosaic virus and bacterial blight.[7]

Over a two-year period (from 1971 to 1973), Hahn drew on the genes from the Ceara rubber × cassava hybrid (58308) and developed real cassava varieties that were resistant to the mosaic virus.[8] Hahn then set about developing mosaic-resistant, high-yielding varieties by crossing mosaic-resistant varieties with many other high-yielding varieties from West Africa and Brazil and selecting and testing clones at the farm level in different agro-ecologies (Hahn, Howland, and Terry 1980; Otoo et al. 1994; Mba and Dixon 1998).

From 1973 to 1977, the IITA cassava program established a partnership with the Shell BP Petroleum Development Company of Nigeria Limited (Shell-BP) in a high rainforest village in the delta area in Nigeria where Shell-BP was producing oil. Shell-BP hired an agronomist and launched a development program to assist cassava farmers in the area. In 1974, IITA scientists conducted a diagnostic survey and found that severe bacterial blight infection and low yield were the main cassava production problems in the area. In collaboration with Shell-BP, IITA conducted on-farm testing of the IITA's clones to select varieties for mosaic disease and bacterial blight resistance, high yield, and root quality.

After six years (1971 to 1977) of research, Hahn and his staff achieved the goal of developing high-yielding, mosaic-resistant TMS varieties. The IITA released the new varieties in Nigeria in 1977. These new high-yielding, mosaic-resistant varieties included TMS (Tropical Manioc Selection) 50395, 63397, 30555, 4(2)1425, and 30572 (hereafter cited as TMS varieties). The TMS varieties were then aggressively diffused by the IITA in collaboration with national extension services, the National Root Crops Research Institute (NRCRI), the Cassava Growers Association, the World Bank, the International Fund for Agricultural Development (IFAD), Shell BP, the AGIP Oil Company, the churches, and the mass media.[9]

The COSCA researchers discovered that the farm-level yield of the TMS varieties in Nigeria was 40 percent higher than that of the local

**Table 5.2.** Nigeria: Yield and Yield Components of Local and IITA Improved Varieties, 1991 to 1992.

*Source: COSCA Study.*

| YIELD | LOCAL VARIETIES | | | | | IMPROVED VARIETIES | | | | |
|---|---|---|---|---|---|---|---|---|---|---|
| | MEAN | MIN. | MAX. | ST. DEV. | NO. OF FIELDS | MEAN | MIN. | MAX. | ST. DEV. | NO. OF FIELDS |
| Yield (tons/ha) | 13.41 | 1.25 | 37.10 | 10.02 | 105 | 19.44 | 3.20 | 36.00 | 8.32 | 34 |
| Stand Density (000/ha) | 10.59 | 1.21 | 41.25 | 31.71 | 105 | 9.60 | 1.22 | 20.10 | 17.03 | 34 |
| Avg. Root Wt. (kg/root) | 0.38 | 0.05 | 1.39 | 0.25 | 105 | 0.45 | 0.09 | 1.14 | 0.23 | 34 |
| No. of Roots/plant | 4.03 | 0.37 | 12.19 | 2.01 | 105 | 6.17 | 3.15 | 35.68 | 5.49 | 34 |
| Harvest Index | 0.56 | 0.32 | 0.79 | 0.13 | 105 | 0.59 | 0.30 | 0.71 | 0.14 | 34 |

varieties when grown without fertilizer (table 5.2).[10] COSCA researchers also found that TMS 30572 was the most popular variety grown in Nigeria in the early 1990s.

Hahn and his colleagues have been described as "the quiet revolutionaries" (Wigg 1993). The yield performance of the TMS varieties is comparable to that of the green revolution wheat and rice varieties in Asia in the 1960s and 1970s (Ruttan 2001).

The COSCA researchers found that the TMS varieties were more resistant than local varieties to the mosaic virus, bacterial blight, the mealybug, and the green mite (table 5.3). The TMS varieties attain their peak yield around thirteen to fifteen months after planting, as compared with twenty-two to twenty-four months for local varieties (fig. 5.1). Nevertheless, COSCA researchers found that Nigerian farmers who planted the TMS varieties and sold *gari* to urban consumers desired varieties that could be harvested less than twelve months after planting without yield loss in order to be able to plant cassava on the same field every year.

Since the TMS varieties are mostly branching types with large canopies, they are good for weed control and have a high leaf yield. The COSCA study found that 50 percent of the area planted to TMS varieties and 55 percent of the area planted to local varieties in Nigeria were intercropped with yam, maize, and the like.[11]

**Table 5.3.** Nigeria: Incidences and Symptom Severity Scores (1–4 scale) of Mealybug, Green Mite, Mosaic Virus, and Bacteria Blight Problems by Local and Improved Cassava Varieties. *Source: COSCA Study.*

| PROBLEM | | LOCAL VARIETIES ($N = 93$) | IMPROVED VARIETIES ($N = 49$) | T-RATIO[a] |
|---|---|---|---|---|
| Mealybug | Percentage infested | 50 | 20 | -- |
| | Mean severity | 2.0 | 1.2 | 3.15 |
| Green Mite | Percentage infested | 26 | 4 | -- |
| | Mean severity | 1.5 | 1.0 | 3.68 |
| Mosaic Virus | Percentage affected | 62 | 73 | -- |
| | Mean severity | 1.9 | 1.5 | 2.45 |
| Bacteria Blight | Percentage affected | 63 | 71 | -- |
| | Mean severity | 1.9 | 1.3 | 4.20 |

[a] indicates $P \leq 0.001$ mean severity is the mean score for plants with symptoms (score of 1 or above).

The Nigerian farmers complained that harvesting the high-yielding TMS varieties by hand was laborious. Southwestern Nigerian farmers who planted the TMS 30572 to produce *gari* for sale in Lagos reported that they had to cut back drastically on the area planted to cassava because they lacked enough seasonal labor to harvest and process the crop in a timely fashion.

The COSCA study farmers in Nigeria who produced cassava as a cash crop and made *gari* for sale to urban consumers praised the TMS varieties as being ideal for *gari* production. However, they complained that peeling the TMS varieties was laborious and resulted in substantial waste because the roots could be peeled only by slashing the skin and part of the root-flesh with a sharp knife. Sometimes the peelers were injured while peeling.

Mechanized machines have not been developed for cassava peeling because cassava roots vary in size and shape. For example, the COSCA farm-level yield measurements show that the roots of the TMS varieties varied in shape and ranged in size from 0.10 kilogram to 1.14 kilogram per root. There is a need for breeders and engineers to work together to develop varieties that produce roots with uniform shape and size, and to develop mechanized peeling machines.

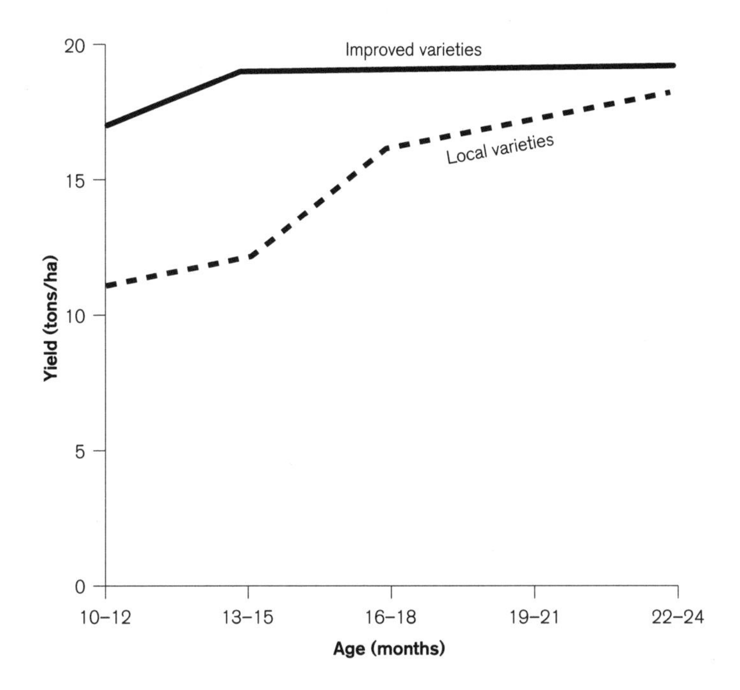


**Figure 5.1.** Nigeria: Cassava Yield by Age for Local and Improved Varieties, 1992. *Source: COSCA Study.*

In Nigeria, the COSCA study found that the roots of the TMS varieties were lower in cyanogen content (an average of 2.20 on a scale of 1 to 3) than those of local varieties (an average of 2.35).[12] Sweet cassava occupied roughly 30 percent and bitter cassava about 70 percent of the area planted with the TMS varieties, the same proportion as for the local varieties.

The rapid rate of diffusion of the TMS varieties in Nigeria meant that the new varieties were displacing local varieties. To protect the genetic diversity in cassava in Africa, Hahn proceeded to explore and conserve the African cassava genetic materials. For example, by 1989 a collection of around fifteen hundred cassava clones was maintained at the IITA headquarters in Ibadan, Nigeria (Ng 1992). Subsequent genetic resources exploration and conservation programs have included the collection of more African local varieties, a survey of the needs of national programs, and introducion of exotic germplasm from Latin America through CIAT (based in Columbia) to Africa (Hahn, Howland, and Terry 1980).

### ■ Building National Cassava Research Capacity

Dr. Hahn realized that national cassava research programs were underfunded and in a state of disarray. Most national agricultural research systems in Africa did not have cassava research programs in the early 1970s. Hahn realized that IITA needed to help develop strong national cassava research programs in cassava-producing countries in Africa in order for IITA's cassava varieties and agronomic practices to be evaluated over a wide range of African agro-ecologies. Hahn took steps to persuade the governments of several of the large cassava-producing countries to establish cassava research programs in the agricultural research systems.[13] For example, a rapid distribution of the IITA's clones to national programs facilitated the development of the TMS varieties (Beck 1980).[14]

To facilitate cooperation among scientists in national programs, the International Society for Tropical Root Crops–African Branch (ISTRC–AB) was founded in 1978. Regional root crop research networks were set up in Eastern and Southern Africa in 1987–88 with support from the IITA.

National research centers in Africa rarely have enough scientists to constitute effective cassava breeding programs. For example, in Ghana in the year 2000, there were only three full-time and a few part-time scientists in the cassava research program of the Crops Research Institute, Kumasi, the national institute responsible for cassava research (Otoo 2000). By contrast, in Brazil there were twenty full-time scientists working on cassava in the Embarapa Cassava and Fruit Crop research institute in the year 2000 (Matos 2001).

Realizing that the dearth of trained scientific manpower would be a major obstacle to the breeding, on-farm testing, and diffusion of the TMS varieties, Hahn quickly established a training program at IITA for national cassava scientists. By 1994, when Hahn retired, 25 national cassava scientists had been trained at the Ph.D. level, 43 had been trained at the M.S. level, and 1,165 had participated in IITA group training programs.

Yet the issue of incentives is another vexing problem. Scientists in

*Dr. S. K. Hahn, leader of the IITA's Cassava research program with the new high-yielding TMS varieties.*

*Plant cuttings of the IITA's TMS varieties being prepared for distribution to farmers.*

*There was often a scramble for the plant cuttings of the TMS varieties when the IITA's distribution team stopped in villages.*

*Cassava with green mite (left) and the IITA's TMS varieties which are resistant to the green mite (right).*

*The TMS varieties gave 40 percent higher yields (right) without fertilizer than the local varieties (left).*

*Researchers of COSCA interviewing cassava farmers in the Congo.*

*Felix Nweke, COSCA research team leader, in a field of the TMS varieties in Nigeria.*

most national programs in Africa are poorly motivated to engage in scientific research. In fact, they are often treated like second-class clerks. Dr John Otoo, the leader of Ghana's cassava program, recently reported that the salaries of his staff were "too low to quote" (Otoo 2000).[15] Ghana's Crop Research Institute had only four computers, with sporadic connections to the internet. Today, it takes an average of two weeks to get an email response from John Otoo. By contrast, in 1999 every scientist at the Embarapa Cassava and Fruit Crop Research Institute in Brazil had a computer connected to the Internet on his or her desk .

In Nigeria, salaries for all government workers were raised in the year 2000. The salaries and allowances of scientists in research institutes and universities were increased from an average of US$100 equivalent per month to US$1000 equivalent per month. Yet the scientists still were not provided with an adequate operating budget. For example, the operating budget of the postharvest unit of the National Root Crops Research Institute in Nigeria was equivalent to US$38 for the month of January 2001 (Oti and Asumugha 2001).

The lack of incentives and income to meet basic needs has forced national scientists to engage in a variety of other activities to make ends meet. COSCA researchers discovered that livestock (poultry, sheep, and goat) keeping was a common secondary occupation of African scientists and academicians. For example, in 1989, one professor at Makerere University in Uganda kept sixty sheep and goats in his university housing compound. To encourage the national researchers in the six countries to carry out the COSCA study, they were paid a monthly stipend of US$200 from COSCA in addition to their university or government salaries.

### Insights and Research Challenges

The long time period required to develop scientific capacity within Africa is one of the major lessons that emerges from the COSCA study. Unfortunately, many policy makers expect breeders to develop high-yielding varieties in an unrealistic period of time. This chapter has shown that it has taken more than forty years (from 1935 to 1977) of hard work to develop the TMS varieties. The evolution of cassava breeding in Africa can be described as a human ladder. Starting in the 1930s, each generation of breeders climbed on the shoulders of the past generation until they hit the jackpot with the development of the TMS varieties in the mid-1970s (table 5.4).

Storey and Nichols were among the first generation of cassava researchers in Africa. They started in 1935 at the Amani research station in Tanzania and developed Ceara rubber × cassava hybrids that were resistant to the mosaic and brown streak diseases but did not stand erect and produced a low root yield that was of poor food quality. Jennings took over the cassava program at the Amani research station in 1951 and developed Ceara rubber × cassava hybrids with wider resistance to the diseases but did not improve on the lodging, low root yield, or poor food quality. The Amani research program was terminated in 1957, and the leadership for cassava research shifted to Nigeria and other countries.[16] When S. K. Hahn joined IITA in 1971, he started cassava research by drawing on the germplasm of Jennings's selections and adding resistance to lodging, earliness of bulking, high yield, and good root quality.

Political turmoil in Africa has frequently undercut the continuity of breeding programs. Nigeria's Civil (Biafra) War (from 1967 to 1970) destroyed twelve years of breeding research by Beck and Ekandem (from 1958 to 1970). The long-term growth cycle of the cassava plant relative to maize, for example, introduces another element of risk. Some cassava plants are ready for crossing in at five months after planting, but several varieties do not flower regularly because they are sensitive to weather. This means that a breeding program can lose a year or more when a breeding stock fails to flower in a particular year because of unfavorable weather conditions.

What is the future research agenda? The improvement of the cassava genetic resource pool represents a work in progress. A. G. O. Dixon joined IITA as a cassava breeder in 1989 and assumed the leadership of IITA's cassava improvement program when S. K. Hahn retired in 1994. Dixon noted that in spite of the excellent progress in cassava breeding made by the IITA's cassava program from 1971 to 1994, several problems require urgent attention. These are the green mite, the cassava anthracnose

**Table 5.4. The Human Research Ladder: Linkages between Cassava Research in Tanzania and Nigeria, 1935–2000.**

| STATION | YEAR | RESEARCHER | GOAL / APPROACH | ACHIEVEMENT / COMMENT |
|---|---|---|---|---|
| **TANZANIA, 1935–1957** | | | | |
| Amani | 1935–37 | Storey | Control cassava mosaic virus and cassava bacterial blight by introduction of resistant varieties. | Over 100 cassava varieties were introduced from various parts of the world. |
| | 1937–51 | Storey and Nichols | Rubber tree species × cassava hybridization. | Developed Rubber tree × cassava hybrids resistant to mosaic. |
| | 1951–57 | Storey and Jennings | Intercross the Storey/Nichol's hybrids to increase resistance to mosaic disease. | Developed hybrids that showed higher and more stable resistance over a wide area (including the Ceara rubber × cassava hybrid, 58308) than the hybrids created by Storey and Nichols and distributed them to several countries, including Nigeria. The Amani research program was terminated in 1957. |
| **NIGERIA, 1958–2000** | | | | |
| Moor Plantation | 1958–60 | Beck and Ekandem | Combine the mosaic disease resistance genes of the Ceara rubber × cassava hybrid, 58308, with the genes for high yield from West African varieties. | Program was transferred to Umudike research station at independence in 1960. |
| Umudike | 1960–70 | Ekandem | Combine the mosaic disease resistance genes of the Ceara rubber × cassava hybrid, 58308, with the genes for high yield from West African varieties. | The breeding work at Umudike was terminated because of Nigerian Civil War (1967–70). |
| IITA, Ibadan | 1971–94 | Hahn | Combine the mosaic-resistant genes of the Ceara rubber × cassava hybrid, 58308, with genes from local and exotic varieties with high yield, good root quality, low cyanogen content, and resistance to lodging. | The famous TMS varieties were developed from 1971 to 1977 and released to farmers in Nigeria in 1977. |
| | 1994–00 | Dixon | Enhance earliness of bulking and root carotene, develop appropriate canopy sizes for leaf harvest and intercropping, and improve pest and disease resistance. | Developed varieties that are resistant to more pests and diseases and have different canopy sizes and distributed them to national cassava research programs in Africa for on-farm testing and selection. |

disease; and the mosaic disease. The cassava mosaic epidemic, which began in Uganda in the early 1990s, is rapidly spreading in Western Kenya and toward the Lake Victoria zone of Tanzania (Legg 1998).

Dixon's breeding goals are to develop varieties that are resistant to multiple pests and diseases, to enhance earliness of bulking and carotene content in the root, and to develop varieties with appropriate canopies for leaf harvest and intercropping (Dixon 2000). Dixon has developed varieties that are resistant to multiple pests and diseases and have different canopy sizes and has distributed them to national cassava research programs in Africa for on-farm testing and selection (Dixon 2001).

Commercial cassava producers in Nigeria and Ghana desire varieties that will attain a maximum yield in fewer than twelve months so that they can grow cassava on the same field for a number of years in succession. Dixon's goal of developing varieties that can be harvested at fewer than twelve months after planting without a loss in yield is designed to help these commercial farmers.

Farmers in Nigeria who plant the TMS varieties and make *gari* for sale in urban centers are facing serious labor bottlenecks at the harvesting and peeling stages. The TMS varieties are not suitable for the development of mechanized harvesting and peeling because their roots lack uniformity in shape and size. In order for the cassava transformation to continue, a solution must be found to the labor constraints faced by commercial cassava farmers. Addressing the problem of labor constraints and reducing the bulking period of cassava roots will improve the productivity of the cassava system, raise farm incomes, and reduce cassava prices to consumers. Without question, developing early-bulking varieties that are suitable for mechanized harvesting and peeling is a critical research challenge over the coming years.

## Summary

Cassava genetic improvement research has been conducted by African smallholders, by researchers in colonial and postindependence research stations, and by scientists in international research centers such as the

IITA and the CIAT. Farmers selected self-seeded plants from local varieties that possessed the attributes which they desire, such as high yield, early bulking, in-ground storage, pest and disease tolerance, processing qualities, large canopy, and low cyanogen level.

Research on cassava by colonial governments was inspired by the appearance and rapid spread of the mosaic disease in East Africa and by the colonial governments' determination to encourage the planting of cassava as a famine-reserve crop in their colonies. The colonial governments set up cassava research programs in various countries in West, Central, and East Africa, but the most successful was the one set up at the Amani research station in Tanzania in the 1930s under the direction of a brilliant but unheralded researcher, H. H. Storey. The mosaic disease-resistant hybrids developed by Storey and his associates at the Amani research station between 1935 and 1957 served as the foundation for IITA's research program on cassava in the 1970s and 1980s.

Soon after most African countries became independent in the 1960s, cassava research began to languish in most national research programs. In 1971, IITA established a research program on cassava. The IITA's program built on local materials and on materials collected from various parts of the world (particularly from the Amani research station); helped national agricultural research systems to establish cassava programs; and established partnerships with farmers, the private sector, and numerous national cassava programs. These research partnerships made it possible for the IITA's clones and field practices to be evaluated over a wide range of environments in Africa.

Under the outstanding leadership of S. K. Hahn in the 1970s, IITA developed high-yielding varieties (TMS varieties) in six years (from 1971 to 1977) of research. The new TMS varieties outyielded local varieties on farmers' fields by 40 percent without fertilizer. COSCA studies in Nigeria found that the TMS varieties are superior to local varieties in terms of yield, earliness of bulking, and pest and disease tolerance, and they are as good as the local varieties in terms of their postharvest attributes and for intercropping. Despite these impressive achievements, however, the cassava revolution is still one of Africa's best-kept secrets!

Although the TMS varieties bulk earlier than do the local varieties, they are not suitable for intensive commercial production because they require thirteen to fifteen months to attain maximum bulking. If farmers plant cassava under continuous cultivation, it will be harvested at fewer than twelve months and before it has attained its maximum yield.

Nigerian farmers also complain that high-yielding TMS varieties are laborious to harvest and peel by hand. There is an urgent need for breeders to develop varieties that attain a maximum yield in fewer than twelve months and have roots that are uniform in shape and size in order to facilitate the development of mechanical harvesters and peelers. The COSCA study found that Nigerian farmers who plant the TMS varieties and produce *gari* for urban markets are often unable to grow as much cassava as they want because of shortage of labor for harvesting and processing the crop.

To summarize, although major progress has been achieved in developing mosaic-resistant and high-yielding cassava varieties, farmers are demanding mechanized techniques for harvesting and processing these varieties. We shall discuss these complex issues in more detail in chapters 6 and 8.

# Agronomic Practices

## Introduction

We have shown in chapter 5 that the development of high-yielding cassava varieties represents a powerful but incomplete engine of growth of the cassava industry. Ultimately, the adoption and spread of high-yielding varieties will be influenced by the development of improved agronomic practices and labor-saving harvesting technology. This chapter draws on the COSCA findings to discuss the evolution of the following agronomic practices: length of fallow period; quality of planting material, plant density (plant spacing) and planting date; cropping pattern (intercropping or mono-cropping); labor use; and cassava harvesting. The analyses in this chapter will show that farmers adopt profitable agronomic practices that save labor. The chapter also shows that progressive farmers who are adopting improved agronomic practices need labor-saving harvesting technology because production and harvesting labor bottlenecks are limiting the amount of cassava farmers are able to produce.

### Three Types of Fallow Systems

There are three main types of fallow systems in Africa. The first is long fallow and it refers to fewer than ten years of continuous cultivation followed by ten or more years of fallow. The second type, short fallow, is fewer than ten years of continuous cultivation followed by fewer than ten years of fallow between crops. The third type, continuous cultivation, refers to at least ten years of continuous cropping. An average of 4 percent of the cassava fields surveyed in the six COSCA study countries were under the long fallow system, 79 percent were under short fallow, and 17 percent were under continuous cultivation (table 6.1).

#### ■ Long Fallow System

When population density was low, long fallow was common and there was no guarantee that farmers would return to the original farmed area after a definite period of time (Okigbo 1984). The COSCA study found that cassava cultivation under the long fallow system has declined in all six study countries because of population growth. In each COSCA study country, farmers reported that they believed that most of their fields would recover soil fertility in ten years or fewer of fallow.

#### ■ Short Fallow System

Data from the six COSCA study countries show that farmers produce cassava under short-fallow system for a variety of reasons, among them are cassava's long growth period, pest and disease problems, and compatibility with crops grown in association with cassava. Pest and disease infestations increase with repeated cultivation of cassava in the same field in most environments. The COSCA study found that under high population pressure on land, farmers produce cassava under short fallow in order to break the pest and disease cycles when the problems are serious.

For example, Kazimzumbwi village in the coastal strip along the Indian Ocean in Tanzania is characterized by a high population density (ninety-six persons per square kilometer) and good road access to markets where cassava is sold. Yet farmers in the Kazimzumbwi village practice cassava production under the short fallow system because of

**Table 6.1.** Percentage of Fields of Different Crops under Different Fallow System: Average for Six COSCA Study Countries. *Source: COSCA Study.*

| CROP | LONG FALLOW | SHORT FALLOW | CONTINUOUS CULTIVATION | TOTAL |
|---|---|---|---|---|
| | | PERCENTAGE | | |
| Cassava | 4 | 79 | 17 | 100 |
| Yam | 14 | 82 | 4 | 100 |
| Rice | 2 | 69 | 29 | 100 |
| Beans (or peas) | 10 | 66 | 24 | 100 |
| Sweet potato | 0 | 62 | 38 | 100 |
| Average | 5 | 75 | 20 | 100 |

serious problems with the cassava mosaic disease and green mite. Cassava is intercropped with beans and maize. The three crops are planted in November, with maize being harvested three months after planting, beans six months after planting, and cassava twelve months after planting because of high market demand due to high population density and easy access to markets. The field is left in fallow for twelve months to break the cycles of the mosaic disease and green mite.

In West Africa, cassava is often grown in association with yam, which requires fertile soil. Farmers in high population density areas who grow cassava in association with yam adopt a short fallow system.[1] Cassava production in Nimbo, a village in Eastern Nigeria, is an example of cassava and yam production under short fallow. The village is characterized by high population density (one hundred persons per square kilometer); a market center that is visited regularly by cassava middlemen (traders and processors); and mild attacks of cassava pests and diseases.

In Nimbo, yam, maize, and melon are planted in April and cassava is planted in the same field in June. The cassava is planted two months after the other crops in order to maintain a cassava stand density in the intercrop while at the same time reducing competition for insolation and soil fertility between cassava and the yam. Yam, which has a seven-month growth period, begins senescence before cassava develops a canopy. Cassava is harvested at twelve months after planting and immediately replanted because of high population density.[2] After harvesting the second crop, farmers leave the field to fallow for three years.

Although cassava may be harvested as few as six months after planting, most of the varieties cultivated in Africa do not attain a maximum yield until twenty-two to twenty-four months after planting. Therefore, farmers in high population density areas who harvest cassava under the short fallow system at twelve months or fewer after planting do not get a maximum yield.

#### ■ Continuous Cultivation

Cassava production under continuous cultivation is a response to increasing population pressure and the need to intensify land use in many African countries. Cassava production in the village of Magamaga, which is located thirteen kilometers outside of the capital city of Kampala in Uganda, is an example of cassava production under continuous cultivation. Banana is the main food staple in this area, and cassava is the secondary staple. The village of Magamaga is characterized by mild problems with cassava pests and diseases, high population density (192 persons per square kilometer) and proximity to an urban market center, Kampala, which is accessible by motor vehicle. Cassava is intercropped with maize, beans (or peas), millet, and sesame. Cassava is planted in March each year and harvested at eight months after planting. The field is replanted to cassava after only four months of fallow. Beans (or peas) are harvested two months and maize four months after planting. This cropping pattern is repeated in the same field every year for at least ten years in a row. As the length of fallow declines under increasing population pressure on land, farmers in Magamaga village are using cassava to replace banana and other crops in the cropping system. In Uganda, farmers in several COSCA study villages reported that they were replacing banana with cassava in their cropping system because of declining soil fertility.

### Cassava Planting

We shall now turn to a discussion of cassava planting material, stand density, planting date, and intercropping. The COSCA study found that cassava is planted more frequently on flat seed-bed than any other major food crop except rice.[3] In well-drained soils, cassava is normally grown on flat seed-bed, but in poorly drained soils, cassava is grown on mounds or ridges to enhance soil aeration (Hahn 1984). The size of the mound or the ridge varies depending on soil drainage conditions. Cassava is planted in a staggered form, almost at random, when it is on flat seed-bed. Mounds, too, are often made in staggered formation. When planting is on ridges, however, cassava is usually planted in rows.

## ■ Planting Material

Cassava is vegetatively propagated from stem cuttings. The COSCA study found that the main sources of planting materials were the farmers' fields, neighbors, and sometimes cassava marketing middlemen. Some COSCA farmers who produce cassava as a cash crop plant healthy-looking cuttings from plants not older than twelve months. Yet, not all farmers discard stems affected by pests or diseases, even though replanting these stems is a common way of spreading those pests and diseases (Rossel, Changa, and Atiri 1994).

In the forest zone, where biomass production is higher than in the savanna zone, there is an abundance of cassava stem cuttings for planting. In the savanna zone, however, cuttings for planting are often in short supply. The COSCA study found that farmers in the savanna zone discontinue the use of varieties that produce low yields of the planting material. The multiplication rate of cassava planting material is low in comparison with that of crops, such as grains, which are propagated by seed. Cassava stem cuttings are bulky, highly perishable, and dry up within a few days after they have been harvested.

## ■ Plant Density

Cassava plant density is an important agronomic consideration because, as we have shown in chapter 4, it is positively and strongly correlated with the cassava yield. The most commonly recommended spacing for cassava is 1 m × 1 m, which is equivalent to a plant density of ten thousand stands per hectare (Onwueme 1978). Yet the average farm-level plant density for the six COSCA study countries was eight thousand plants per hectare, with a range of from five hundred plants per hectare to forty thousand plants per hectare. The COSCA study found that the cassava stand density in the farmers' fields varies widely depending on climatic zone, cassava morphological characteristics such as branching type and leaf shape, soil fertility status, seed-bed type, cropping pattern, and so on. Farmers who plant cassava as a cash crop plant at higher plant densities than do farmers who plant the crop as a famine-reserve crop or as a rural food staple.

The average cassava plant density varies by country (fig. 6.1). The low plant density in Uganda is due to forest and midaltitude environments; the high plant density in the Congo is due to low soil fertility; and the relatively high plant densities in Nigeria and Ghana are due to production of cassava as a cash crop. In Nigeria and Ghana, the cassava stand density is higher in fields in which cassava is planted for sale (8,200 plants per hectare) than in fields in which cassava is planted for home consumption (7,600 plants per hectare). The mean cassava plant density is also higher in high population density zones (8,800 plants per hectare) than in low population density zones (7,000 plants per hectare) because of population pressure on land.

The factors affecting cassava plant density are so numerous that the optimal plant density for high cassava yield is location specific. Local agronomic research is needed to determine the optimum cassava plant density. Yet research on cassava is not carried out at the necessary local levels in any of the six COSCA study countries because the National Agricultural Research Systems (NARs) do not have the resources necessary for detailed cassava field experiments.

## ■ Planting Date

Cassava does not have critical planting date, as long as there is enough moisture at planting for rooting to commence. In West Africa, the rainy season normally starts by February–March and ends by October–November. Most food crops, except cassava, are planted at the onset of the rainy season. The COSCA study shows that cassava is planted from the beginning of the rainy season to the end of the rainy season in West Africa.

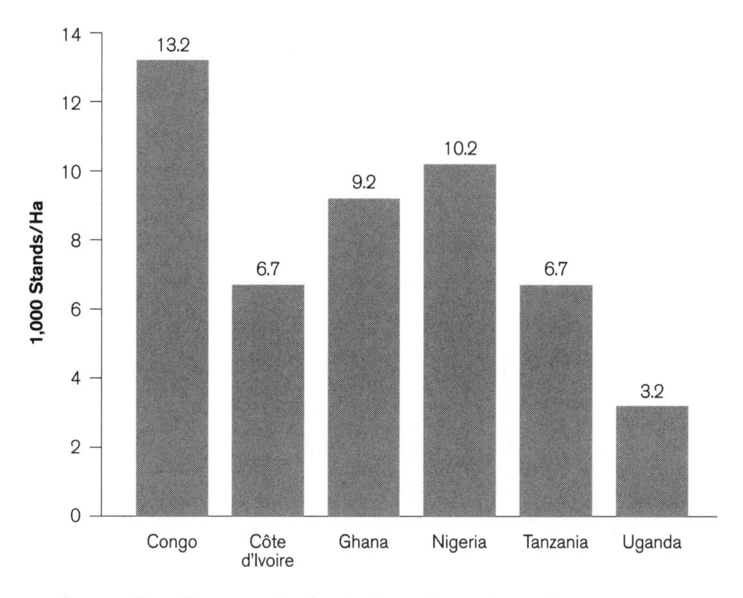


**Figure 6.1. Cassava Plant Density in the Congo, Côte d'Ivoire, Ghana, Nigeria, Tanzania, and Uganda.**
*Source: COSCA Study.*

The cassava planting date is also significantly influenced by the intercropping needs of farmers. The COSCA study found that the farmers delay the planting of cassava in intercropped fields, depending upon the growth cycle of the associated crop. If the associated crop is a short-cycle crop such as maize, cassava is often planted one month later than the first crop. This means that in some cases maize and cassava are relay-cropped (planting cassava as maize is about to be harvested) because maize will then be harvested before cassava establishes a canopy. In these cases, competition for insolation and soil fertility between the two crops is minimized, if not eliminated, thus allowing the monocrop stand density to be maintained for cassava. If the associated crop, such as yam, has a long growing period, cassava is often planted as much as three months later than the first crop.

■ Intercropping

In the six COSCA study countries, the cassava yield is higher when the crop is grown in mono-culture (14 tons per hectare) than when it is intercropped (11 tons per hectare). Yet cassava is frequently intercropped (Fresco 1986). In the six COSCA study countries, 60 percent of the cassava fields are intercropped and 40 percent are mono-cropped. Cassava/maize intercropping is by far the most common, constituting about 50 percent of all cassava-based intercropped fields. Other common combinations are cassava/bean (or pea), cassava/banana (or plantain), cassava/rice, cassava/millet (or sorghum), cassava/yam, and cassava/sweet potato. Half of the 124 farm households in northern Nigeria studied by Norman (1974) identified high total output as the main reason for intercropping, followed by a shortage of land and a shortage of labor. Cassava's flexible planting schedule, its wide interspacing, and its slow rate of growth relative to maize, for example, make it suitable for intercropping.

## Cassava Production

Labor is the main component of the cost of cassava production. Yet the conventional wisdom is that cassava requires relatively low labor inputs for production (Hendershott et al. 1972; Jones 1959). However, COSCA research confirms that this conventional wisdom is valid only where cassava is produced as a famine-reserve crop or as a rural food staple.

The COSCA study reveals that the average cassava production and harvesting labor varies from 173 person-days per hectare in the Côte d'Ivoire to 201 person-days per hectare in the Congo and 222 person-days per hectare in Nigeria (table 6.2). Labor is used for bush clearing, tillage, planting, weeding, and harvesting.[4] The use of a total of 201 days of labor per hectare in the Congo is due to the large amount of labor required for clearing forests. The labor inputs are high in Nigeria and Ghana because cassava is produced as a cash crop by progressive farmers who are early adopters of better tillage and timely weeding practices.

Farmers in Nigeria and Ghana use more hired labor in cassava production than farmers in the other four COSCA study countries, because

**Table 6.2.** Cassava Production by Task and Days per Ha in the Congo, Côte d'Ivoire, Ghana, Nigeria, Tanzania, and Uganda. *Source: COSCA Study.*

| TASK | CONGO | CÔTE D'IVOIRE | GHANA | NIGERIA | TANZANIA | UGANDA |
|---|---|---|---|---|---|---|
| | | | DAYS PER HA | | | |
| Land clearing | 66 | 53 | 44 | 49 | 54 | 45 |
| Seed-bed preparation | 21 | 29 | 31 | 41 | 27 | 31 |
| Planting | 39 | 22 | 28 | 32 | 27 | 28 |
| Weeding | 27 | 28 | 34 | 38 | 28 | 32 |
| Harvesting | 48 | 44 | 53 | 62 | 46 | 52 |
| TOTAL DAYS | 201 | 173 | 191 | 222 | 182 | 187 |

cassava is grown mostly as a cash crop for urban consumption (fig. 6.2). However, as wage rates increase, it will become difficult for farmers in these countries to continue to produce and harvest cassava at prices competitive with grain. Therefore, improvement of the productivity of the cassava system requires the development of labor-saving methods for cassava production and harvesting tasks.

Labor-saving technologies, such as mechanized equipment, are available for land clearing and tillage, but they are not widely used in cassava production in Africa because available mechanized technologies are not suitable for preparing seed-beds for cassava. For example, in well-drained soils, farmers plant cassava on flat seed-bed with minimum tillage. In poorly drained soils, however, farmers plant cassava on high ridges and mounds to enhance drainage and soil aeration. Yet mechanized technologies are not efficient in making high ridges and mounds in poorly drained soils.[5]

Land clearing is a labor-intensive task under the short-fallow cassava production system in Africa. In the Congo, the difficulty of land clearing limits the amount of cassava a woman can cultivate without male assistance (Fresco 1986). In Nigeria, a land clearing agency, the National Agricultural Land Development Authority (NALDA), clears large tracts of forestland and allocates plots to smallholders for the planting of cassava and other crops.

In the mid-1970s, IITA scientists initiated research on minimum tillage, which is defined as planting on the flat without plowing, ridging,

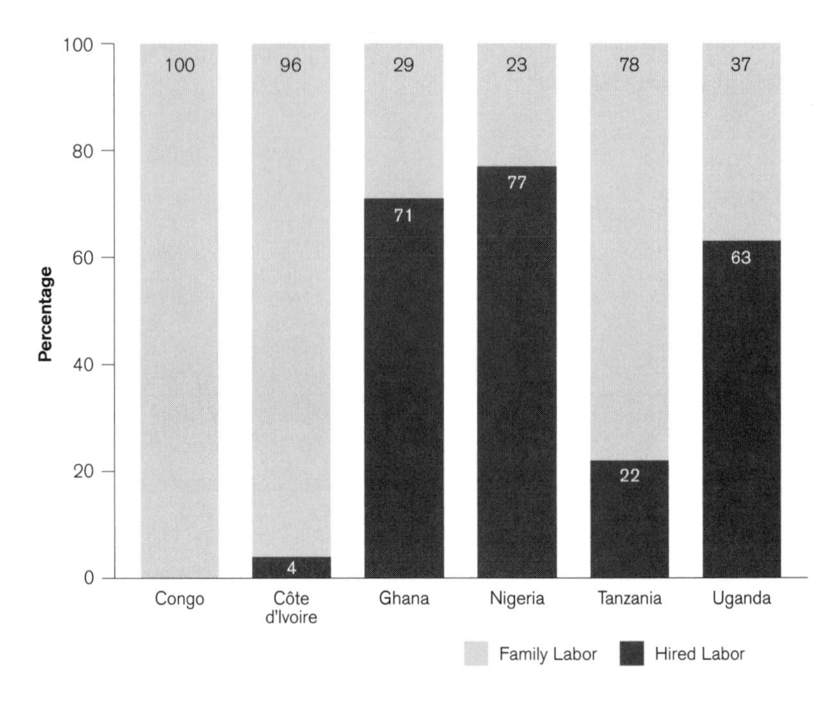


**Figure 6.2.** Percentage of Cassava Fields Using Hired Labor in Six COSCA-Study Countries. *Source: COSCA Study.*

or mounding, as an alternative to tillage (plowing, ridging, or mounding) in order to reduce plowing labor, prevent soil erosion, and maintain soil structure. The studies show that although the minimum-tillage method controlled soil erosion and maintained soil structure, it required a chemical herbicide for weed control (International Institute of Tropical Agriculture 1992b). Experiments at the IITA station in Ibadan to determine the effect of the minimum-tillage practice on cassava yield revealed that the yield was significantly higher when cassava was planted on ridges, because of better weed control (Akobundu 1984; Ezumah, Ohuyon, and Kalabari 1984). Yet the minimum-tillage practice has not been adopted by farmers, because it reduces yield and farmers generally do not have access to chemical herbicides.

Cassava planting labor is defined here to include labor used to collect the planting material from the farmer's previous fields and to plant. If the cassava farm size per household increases, labor for the cassava planting operation will become a bottleneck because the cassava planting sets are bulky and perishable and they need to be planted as soon as they are harvested. In parts of Nigeria where cassava is produced as a cash crop, a market is developing for cassava planting material.

In all six COSCA study countries, weeding was found to account for a large proportion of total labor input. Although some agricultural scientists report that cassava requires little weeding when planted in optimal plant populations (Okigbo 1980; Onwueme and Sinha 1991), the COSCA study found that weeds constitute a serious problem in cassava production because generally it takes about two to four months before cassava leaves close the canopy and suppress weed growth (Dahniya and Jalloh 1998).

Mechanized weeding can be adapted for cassava fields but the adaptation will require significant modification in cassava agronomic practices. For example, the mound seed-bed will have to be replaced with a ridge seed-bed and cassava will have to be planted in rows. Smallholders will make these changes only if they are profitable.

## Cassava Harvesting

Most cassava varieties form edible roots within the first six months of planting and may be harvested at that age. If not harvested, the plant keeps growing and the roots continue to bulk (increase in size), in most cases, for up to twenty-four months after planting. Yet even after cassava reaches maximum bulking, harvesting can be deferred until the need arises or to a time convenient for the farmer.

In Africa, harvesting of a field of cassava is generally spread over a period of several months or even up to three to four years. Farmers do not harvest a cassava field systematically from one corner to the other. Rather, a farmer harvests particular cassava stands, depending on variety, size, and location in the field. Farmers also frequently *milk* their cassava plants; that is, they harvest some but not all of the roots of a plant at one time, only to come back later, sometimes after several weeks or even months, to harvest the remaining roots. This practice is more common in countries such as Tanzania, where farmers produce cassava as a famine-reserve crop. Such farmers store their crops in the field and take only as many roots as are needed for a few meals (Kapinga 1995). The cassava-harvesting schedule is influenced by the general food supply outlook, market conditions, and labor availability for harvesting and processing. During drought and periods of attractive market prices farmers will harvest their cassava early, sometimes before roots attain their maximum bulking.

For the six COSCA study countries, the average age of cassava fields at harvest was 11.9 months, with a range of from 6 to 49 months. Among countries that planted mostly sweet cassava varieties (the Côte d'Ivoire, Ghana, and Uganda), the age of cassava fields at harvest was lowest in Ghana (fig. 6.3). Among countries that planted mostly bitter cassava varieties (the Congo, Nigeria, and Tanzania), the age of cassava fields at harvest was lowest in Nigeria. In Nigeria and Ghana cassava is produced mostly as a cash crop.

Usually, if cassava is harvested after twelve months of planting, the field is no longer weeded. The field is, in effect, under bush fallow, and cassava is harvested by wading into the bush fallow, clearing the bush around each cassava plant with a cutlass, chopping off the plant, pulling (or digging up) its roots, and then cutting the roots off the stump. The farmer then moves on to another plant, later gathering the roots for transportation to the home or to a processing center.

Cassava is planted as a cash crop in Nigeria and Ghana. The amount of cassava harvesting labor per hectare was thus higher in Ghana and Nigeria than in the other COSCA study countries (fig. 6.4). In fact, table 6.2 shows that harvesting cassava is the most labor-intensive field task in Ghana and particularly in Nigeria, where the TMS varieties and improved agronomic practices have boosted yields by 40 percent. In Nigeria and Ghana, the labor constraint has shifted from cassava weeding to cassava harvesting. Harvesting is now proving to be a serious constraint on the spread of TMS varieties because labor for cassava harvesting increases in direct proportion to yield. It is not surprising that farmers who plant TMS varieties in Nigeria have sometimes had to suspend planting because they

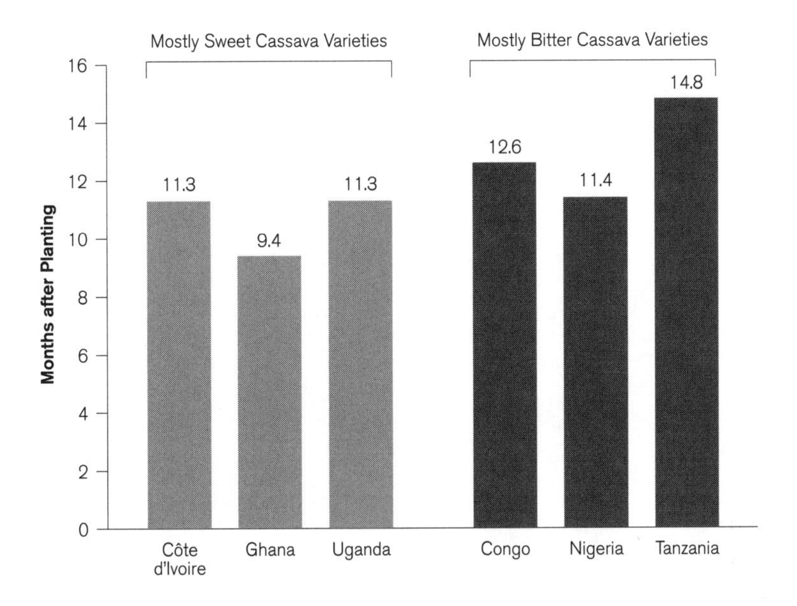


**Figure 6.3. Age of Cassava Fields at Harvest in Months after Planting in the Congo, Côte d'Ivoire, Ghana, Nigeria, Tanzania, and Uganda.** *Source: COSCA Study.*

have been unable to hire sufficient labor to harvest previously planted cassava fields.

To date, most public sector research on cassava has focused on genetic research. Very little research has focused on developing machines to harvest cassava. In Nigeria, the harvesting constraint for cassava is now at a similar stage to that for grain harvesting in the United States at the beginning of the nineteenth century, when grain was still harvested by the same method that had been used in the fourteenth century (Johnson 2000, 6). The invention of the reaper in America in the second quarter of the nineteenth century sharply reduced labor inputs in grain harvesting. The combine then replaced the reaper, and the direct labor inputs required to produce a ton of grain declined by 70 percent in the nineteenth century (Johnson 2000). Without question, a mechanical revolution is now needed to break the labor bottleneck in cassava harvesting among farmers in Nigeria and Ghana who are planting the TMS varieties.

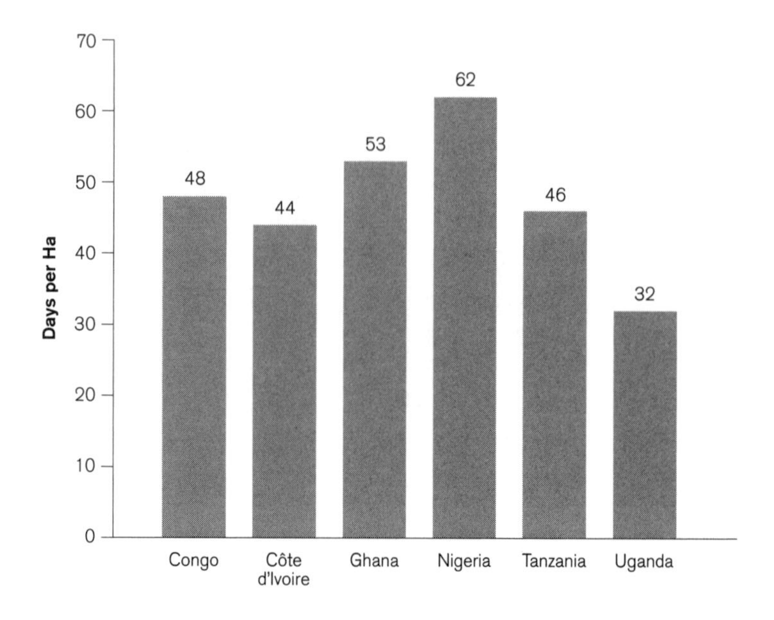


**Figure 6.4.** Number of Person-Days per Hectare Used in Cassava Harvesting in Six COSCA-Study Countries.
*Source: COSCA Study.*

Mechanization of the harvesting operation is more urgent than mechanization of any of the preharvesting tasks because it will facilitate the adoption of agronomic measures that can raise cassava yields. Given the complexity of smallholder production systems, researchers should focus on mechanizing one harvesting task at a time. In the United States, for example, the combine harvester for grain was developed in four stages: the reaper, the binder, the thresher, and finally the combine, in that order (Johnson 2000). The same multistep process should be followed in Africa.

From the mid 1970s to the early 1980s, attempts were made at IITA to adapt mechanized potato harvesters for cassava harvesting. The research was thwarted, however, by different types of cassava seed-beds, irregular cassava plant spacing, variable canopy size, and by the premature termination of the IITA's research on mechanization in the early 1980s (Garman and Navasero 1982).

The high labor requirements of harvesting and transportation are not obvious under the present ad hoc harvesting strategies. When a few stands of cassava are harvested, it is easy for a woman to carry one head-load to her home at the end of the day's work. Yet mechanized harvesting and transportation will be needed when an entire cassava field has to be harvested and transported in one day. In some areas where cassava is planted as a cash crop and where farm roads are accessible to motorized vehicles, such as in parts of Nigeria, entire cassava fields are harvested manually but transported by motor vehicles to processing centers. In most of the cassava-growing areas of Africa, however, farm roads are so poor that cassava can be transported only by head-load. Therefore, the mechanization of cassava harvesting will require improvement in farm roads to facilitate the movement of the mechanization equipment to the cassava fields and the evacuation of cassava.

## Summary

There are three main fallow systems in Africa: long fallow, short fallow, and continuous cultivation. Because of population growth, cassava cultivation under the long fallow system has declined in all six COSCA study countries. The distribution of cassava fields surveyed by COSCA researchers were as follows: long fallow, 4 percent; short fallow, 79 percent; and continuous cultivation, 17 percent. Most farmers now produce cassava under the short fallow system because of cassava's long growth period, pest and disease problems, and compatibility with crops grown in association with cassava. Yet cassava production under continuous cultivation is an increasing practice in many African countries in response to the increasing population pressure on land.

Cassava plant density is an important consideration for farmers because it is strongly correlated with cassava yields. Farmers who plant cassava as a cash crop plant at higher plant densities than do farmers who plant it as a famine-reserve crop or a rural food staple. The factors affecting cassava plant density are so numerous that the optimal plant density for high cassava yield is location specific.

Cassava is reputed to be a crop that requires relatively low labor inputs for production. However, COSCA research confirms that this conventional wisdom is valid only where cassava is produced as a famine-reserve crop or as a rural food staple. When cassava is produced as a cash crop, farmers use significantly more labor. In most of the COSCA study villages, all the field tasks for cassava production are done manually, using simple implements such as hand hoes and a cutlasses. In Nigeria, where the TMS varieties and improved agronomic practices have boosted yields by 40 percent, the labor constraint has shifted from cassava weeding to cassava harvesting. Harvesting is now proving to be a serious constraint on the spread of TMS varieties, because labor for cassava harvesting increases in direct proportion to yield.

The following conclusions emerge from the analyses in this chapter: farmers will introduce new agronomic practices as cassava production becomes profitable. However, progressive farmers who adopt improved agronomic practices are facing production and harvesting labor bottlenecks which are limiting the amount of cassava they can grow.

Without question, agronomic and genetic-induced productivity changes are important sources of growth in the cassava industry. Yet mechanical technology for cassava production and harvesting is needed to complement improved agronomic practices and high-yielding varieties. The mechanization of harvesting is needed to break labor bottlenecks, increase labor productivity, and encourage farmers to increase the use of TMS varieties.

# Diffusion of TMS Varieties

## Introduction

In chapter 5, we concluded that TMS varieties are superior to local varieties in terms of yield, earliness of bulking, and resistance to pests and diseases. The TMS varieties are also at least as good as the local varieties in terms of cyanogen content, ease of processing, cooking quality of fresh roots, and the quality of processed products such as *gari*. This chapter examines the factors responsible for the rapid diffusion of the TMS varieties in Nigeria and the delayed diffusion of the varieties in Ghana and Uganda. We shall address the puzzle as to why there was a sixteen-year delay between the time that the TMS varieties were released in Nigeria in 1977 and the government of Ghana's official release of the TMS varieties in 1993.

## TMS Diffusion in Nigeria

### ■ Food Policy Environment

The dramatic expansion of Nigeria's oil exports in the 1970s increased the real rate of growth of per capita GNP by 5.3 percent and sparked massive rural to urban migration (Akande 2000). In the 1970s, the government used foreign exchange earnings from petroleum export to help pay for food imports. From 1976 to 1985, the annual per capita rice imports increased by more than 1,500 percent of its 1961 to 1965 level. The imported rice was subsidized for consumers because the Naira was overvalued, and the rice was distributed by the Nigerian National Supply Company Limited, a government agency which sold rice at a uniform price nationwide and absorbed transportation costs.[1] The quantum jump in subsidized rice and wheat imports depressed the price of *gari* and acted as a constraint on the spread of the TMS varieties from the late 1970s to 1985.

In 1985, the Nigerian government banned the importation of wheat, rice, and maize and the export of yam and cassava products. In 1986, Nigeria adopted a structural adjustment program (SAP) that consisted of a number of policy reforms, including the devaluation of the Naira from 1.0 Naira per US$1.0 in 1986 to 4.0 Naira per US$1.0 in 1987 (Akande 2000, 11).[2] The National Seed Service (NSS) multiplied and distributed free stem cuttings of the TMS varieties to farmers. The subsidy on fertilizer was increased from 72 percent to 85 percent of the farm-delivered price and fertilizer use increased from 100,000 tons in 1980 to 518,120 tons in 1990 (Akande 2000, 5). The monetary devaluation in 1986 and the farm input subsidies, including the free distribution of TMS stem cuttings, contributed immensely to the uptake of fertilizer and the diffusion of the TMS varieties.

In 1995, an economic liberalization program was introduced and the ban on food imports and exports was removed. The aims of the new policy were to discourage food imports and encourage production and export of food and raw materials in order to raise farm income. From 1994 to 1998, the aggregate agricultural growth rate was 3.7 percent, which was higher than the estimated 3.0 percent population growth rate (Akande 2000, 15). Without question, the economic liberalization policy has helped increase the diffusion of the TMS varieties, has increased output, and has reduced cassava prices to consumers. The director of the Nigerian Federal Department of Agriculture reported in early 2001 that there was a glut of cassava on the market and that the price of *gari* was low.[3]

### ■ IITA's Role in Jump-Starting the Diffusion of the TMS Varieties

In chapter 5, we explained that at the early stages of the IITA's breeding program in 1973, the IITA scientists tested new cassava varieties at the farm-level in collaboration with extension agents, farmers, and several private-sector agencies in Nigeria. Without doubt, the physical presence of the IITA in Nigeria was influential in eliciting the assistance of non-governmental organizations in the diffusion of the TMS varieties. The IITA, in collaboration with the Bendel State Ministry of Agriculture, conducted a diagnostic survey on behalf of Shell-BP (Shell-BP Petroleum Development Company of Nigeria Limited) in a high rainforest village, Agbarho, in the delta area of Nigeria.[4] The aim of the survey was to design a program to help local farmers increase cassava production. IITA scientists discovered that bacterial blight and low yields were the main production problems facing farmers. In 1973–74, IITA planted several of its new varieties in an on-farm experimental plot in the delta area. In 1975–76, Shell-BP provided cassava planting materials for Nigeria's National Accelerated Food Production Program (NAFPP).[5]

In 1975, IITA's scientists helped Texagri (Texaco Agro-Industries Nigeria Limited) establish a four-hundred-hectare cassava plantation in Southwestern Nigeria to grow the TMS varieties and process the roots as *gari*. From 1988 to 1991, Texagric distributed free planting materials to local farmers. The Nigerian Agip Oil Company Limited also multiplied and supplied TMS planting materials to a large number of farmers, cooperative societies, women's associations, and schools. Other NGOs involved in the production, promotion, and distribution of planting materials of the improved varieties included church groups, schools, universities, the Nigerian Cassava Growers Association, and the mass media.

Dr. S. K. Hahn, the head of the IITA's cassava research program, distributed planting materials of improved varieties through churches and schools (Hahn 1999). Hahn, a Catholic, went to different churches each Sunday dressed in his Yoruba tribal chieftaincy regalia.[6] At the end of the mass, he stood at the church's main door with small bundles of the cuttings of the improved varieties, encouraging members of the congregation, especially women, to take the cuttings and test-plant them in their fields. Hahn also visited numerous schools and encouraged children to take the materials to their parents to plant alongside local varieties.

Hahn often accompanied an IITA driver with a truckload of the cuttings of the TMS varieties, stopping at intervals along the highways and donating small bundles to farmers to test-plant. Hahn reports that there was often a scramble for the cuttings when the truck stopped in villages along the highways. In locations where Hahn did not find farmers, he stopped anyway to plant a few cuttings in cassava fields along the highway.[7] Hahn encouraged cassava farmers to launch the Nigerian Cassava Growers' Association, with membership drawn from all the cassava-producing states of Nigeria. The association helped distribute the TMS varieties throughout Nigeria. Hahn also prepared news releases about the TMS varieties and distributed them to Nigerian newspapers and radio and television stations.

In 1984, Natalie D. Hahn, an IITA social scientist, conducted a farm-level survey in Southwestern Nigeria to find ways to encourage women farmers to adopt the TMS varieties. The women reported that they purchased fields of cassava, harvested the roots, and spent a large percentage of their time preparing *gari* for sale in Ibadan. With support from UNICEF (United Nations Children's Fund), a mechanized *gari*-processing center was established in one of the survey villages near Ibadan. The aim was to reduce women's labor inputs in cassava processing so that they could expand production of the TMS varieties in order to "increase food consumption, reduce malnutrition and hunger" (N. D. Hahn 1985, 199).

We now turn to a discussion of other factors in Nigeria that facilitated the diffusion of the TMS varieties: vigorous government research and extension services; a dense network of rural roads; available technology for *gari* preparation; and the mechanized grater.

## ■ National Research and Extension Services

The rapid diffusion of the TMS varieties in Nigeria was facilitated by the collaboration of the National Root Crops Research Institute (NRCRI), the World Bank, and the International Fund for Agricultural Development

(IFAD); by government revenue from the oil sector; and by availability of low-cost gasoline. With the aid of petroleum revenue, the Nigerian government experimented with alternative extension programs and expanded higher education and agricultural research institutions in the 1970s and 1980s. For example, the adoption of the TMS varieties was promoted by the national extension program under the National Accelerated Food Production Program (NAFPP), which was set up in 1972 to design, test, and transfer technological packages for five crops: rice, maize, sorghum, millet, and wheat. In 1974, cassava was added to the list.

Under the NAFPP, extension agents helped farmers prepare 7 × 39 square meter demonstration plots planted with the TMS varieties side by side with local varieties. At harvest time, a panel of local farmers compared the plots, and if TMS varieties were found to be superior, the TMS demonstration advanced to a second phase, which involved fewer plots of a larger size. Farmers were expected to adopt the TMS varieties if they continued to be superior to the local varieties in the second phase of the demonstration. The NAFPP introduced the TMS varieties to all the cassava-producing areas of Nigeria, making it easy for further diffusion by the farmer-to-farmer method of technology transfer. By 1985, the NAFPP was working with 704,000 farmers in the twelve major cassava-producing states of Nigeria.[8]

In 1974, the World Bank financed the establishment of Agricultural Development Projects (ADPs) in three states in Nigeria and, by 1985, ADPs were in operation in every state of the country. The functions of the ADPs included: construction of roads for input delivery and output evacuation, provision of extension service to farmers, and multiplication and distribution of TMS stem cuttings and seeds of other crops. In 1986, for example, the Oyo State ADP distributed the planting materials of the TMS 30572 varieties to fifty-five thousand farmers in the state. The ADPs in the other cassava-producing states also distributed the planting materials to farmers in their states. Thus, the ADP played a significant role in the diffusion of the TMS varieties in Nigeria.[9]

In 1986, the federal government directed the National Seed Service (NSS) to assist the ADPs in the multiplication of the TMS varieties with the assistance of a US$120 million grant from IFAD. The multiplication and distribution of the TMS planting materials was expensive, because one cassava plant produces an average of four plant cuttings while one maize plant produces about one hundred seeds, and cassava-planting material is bulky and perishable and hence expensive to transport. These factors explain why the multiplication is done in villages close to where the planting materials are needed.

Starting in the 1970s, the NARs collaborated with federal and state extension programs to accelerate the adoption of the TMS varieties. Without question, Nigeria has the largest concentration of agricultural research and agricultural higher education institutions in Africa. In 1997, Nigeria had eighteen agricultural research institutes, three universities of agriculture, and twenty-eight colleges of agriculture in addition to thirty general universities (Idachaba 1998).

Improved rural roads facilitated the transportation of *gari* to urban centers in Nigeria. The quality of rural roads is higher in Nigeria than in other COSCA countries. For example, 49 percent of the COSCA villages in Nigeria had a paved road, compared to 20 percent in the Côte d'Ivoire; 19 percent in Ghana; 17 percent in Uganda; 9 percent in Tanzania; and none in the Congo.

To summarize, the TMS diffusion in Nigeria is an African success story par excellence! In 1989, COSCA researchers found that the TMS varieties were grown in 60 percent of the surveyed villages in the cassava-growing areas of Nigeria (table 7.1). The TMS varieties were grown in both the forest and the savanna zones of Nigeria (fig. 7.1). The TMS 30572 variety was the most popular among farmers, especially among those who process it as *gari* for sale in urban markets.

### ■ Profitability of TMS Varieties in *Gari* Production

The COSCA study found that the profitability of the TMS varieties critically depends upon the type of available grating technology.[10] We shall show in chapter 9 that mechanized graters for *gari* preparation are available in 52 percent of the COSCA villages in Nigeria.[11]

Drawing on COSCA data and the classification by Camara (2000), COSCA farmers in Nigeria can be divided into four categories based on

**Table 7.1.** Percentage of COSCA-Study Villages Where the IITA's High-Yielding TMS Varieties Were Planted in 1989–90. *Source: COSCA Study.*

| COUNTRY | NO FARMERS | FEW FARMERS | MANY FARMERS | MOST FARMERS | TOTAL |
|---|---|---|---|---|---|
| | | | PERCENTAGE | | |
| Congo | 97 | 3 | 0 | 0 | 100 |
| Côte d'Ivoire | 100 | 0 | 0 | 0 | 100 |
| Ghana | 100 | 0 | 0 | 0 | 100 |
| Nigeria | 11 | 30 | 36 | 23 | 100 |
| Tanzania | 50 | 50 | 0 | 0 | 100 |
| Uganda | 85 | 5 | 5 | 5 | 100 |

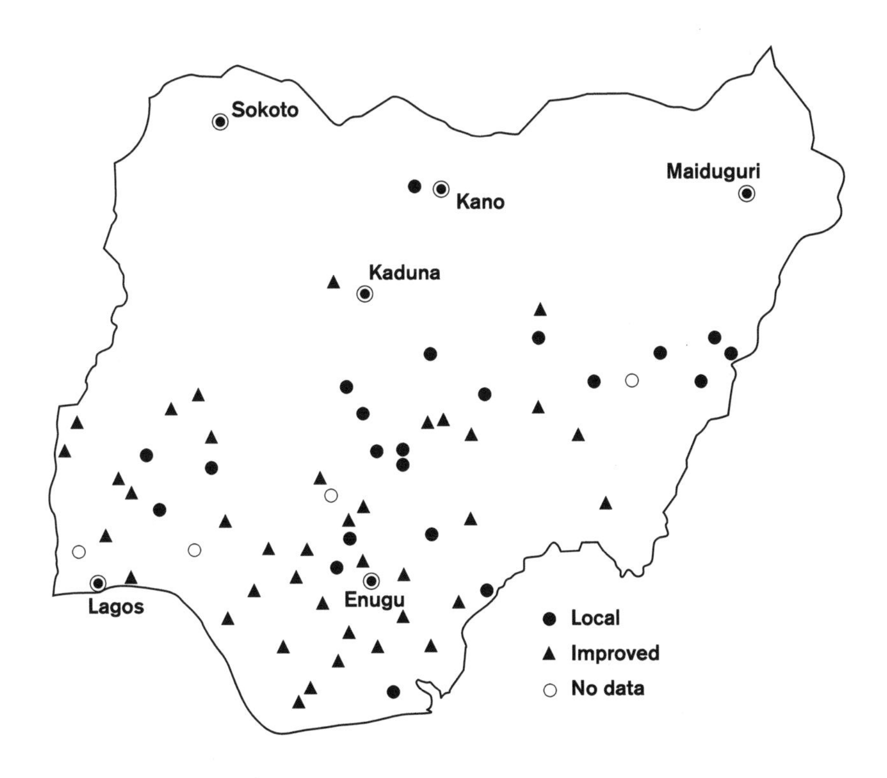


**Figure 7.1.** Nigeria: Location of COSCA-Study Villages Where Farmers Planted the IITA's High-Yielding TMS Varieties in 1989–90. *Source: COSCA Study.*

the variety (local or TMS) planted and the grating method (manual or mechanized) used. The net profit per ton of *gari* is as follows:

- local varieties with manual grating . . . . . . . . . . . . US$2.50
- local varieties with mechanized grating . . . . . . . . US$28.00
- TMS varieties with manual grating . . . . . . . . . . . US$20.00
- TMS varieties with mechanized grating . . . . . . . . US$46.00

Table 7.2 presents a financial analysis of the four combinations of cassava production and *gari* preparation technology. This financial analysis shows that farmers who plant local varieties and grate manually earn a modest net profit of 42 Naira (about US$2.50) per ton of *gari.* Farmers who plant local varieties and use mechanized grating earn 478 Naira (about US$28.00) net profit per ton of *gari* as compared with a net profit of 339 Naira (about US$20.00) per ton of *gari* by farmers using TMS varieties and manual grating. Cassava farmers benefit more from using labor-saving grating technology than from planting TMS varieties. TMS varieties are significantly more profitable, however, when grating is mechanized. For example, farmers planting the TMS varieties and using mechanized grating earned a net profit of 776 Naira (about US$46.00) per ton of *gari.*

The financial analysis shows that the use of labor-saving grating technology is essential for the rapid adoption of TMS varieties. Yet the growing availability of the mechanized grater has shifted the cassava labor bottleneck to harvesting, peeling, and toasting. The COSCA study found that farmers in Nigeria who were growing the TMS varieties frequently reduced the area planted because, owing to labor shortage, they were not able to harvest and process the crop from the previous season's plantings.

Farmers planting the TMS varieties reduce labor cost by importing migrant labor to help in harvesting and rely on local female labor for peeling and toasting. The COSCA study found that among the commercial cassava growers and processors in the Ovia area of Edo State, cassava peeling and toasting were carried out almost exclusively by women, who were paid a lower wage, 15 Naira (US$0.88) per day, than the 21 Naira (US$1.24) daily rate for men in 1991. The women combine the peeling

**Table 7.2.** Nigeria: Financial Budgets for *Gari* Preparation by Alternative Cassava Production and Processing Technologies, 1991. *Source: COSCA Study.*

| BUDGET ITEM | PRODUCTION AND PROCESSING TECHNOLOGIES | | | |
|---|---|---|---|---|
| | LOCAL VARIETIES MANUAL PROCESSING | LOCAL VARIETIES MECHANIZED PROCESSING | TMS VARIETIES MANUAL PROCESSING | TMS VARIETIES MECHANIZED PROCESSING |
| INPUTS/LABOR | | | | |
| Production (man-days/ha) | | | | |
| Bush clearing | 49 | 49 | 49 | 49 |
| Tillage | 41 | 41 | 41 | 41 |
| Planting | 28 | 28 | 28 | 28 |
| Weeding | 34 | 34 | 34 | 34 |
| Subtotal | 152 | 152 | 152 | 152 |
| Harvesting (man-days/ha) | 56 | 56 | 82 | 82 |
| *Total male labor* (man-days/ha) | 208 | 208 | 234 | 234 |
| Processing (woman-days/ha) | | | | |
| Peeling (3.6 woman-days/ton of root) | 39 | 39 | 56 | 56 |
| Grating (9.9 woman-days/ton of root) | 106 | 0 | 154 | 0 |
| Toasting (3.3 woman-days/ton of root) | 35 | 35 | 51 | 51 |
| *Total Female labor* (women-days/ha) | 180 | 74 | 261 | 107 |
| OUTPUTS | | | | |
| Root yield (tons/ha) | 13.41 | 13.41 | 19.44 | 19.44 |
| Useable root yield (80% of root yield) | 10.73 | 10.73 | 15.55 | 15.55 |
| Root-to-*gari* conversion ratio | 0.33 | 0.33 | 0.33 | 0.33 |
| *Gari* yield (tons/ha) | 3.54 | 3.54 | 5.13 | 5.13 |
| Village market price of *gari* (Naira/ton of *gari*) | 3140 | 3140 | 3140 | 3140 |
| COSTS (Naira/ha) | | | | |
| Male labor (21 Naira/man-day) | 4368 | 4368 | 4914 | 4914 |
| Female labor (10 Naira/woman-day) | 2700 | 1110 | 3915 | 1605 |
| Farm transportation (92 Naira/ton of root) | 1233 | 1233 | 1790 | 1790 |
| Grating fee (15 Naira/ton of root) | 0 | 161 | 0 | 233 |
| Bagging (82 Naira/ton of *gari*) | 290 | 290 | 420 | 420 |
| Firewood (207 Naira/ton of *gari*)[a] | 733 | 733 | 1062 | 1062 |
| Transportation to market (235 Naira/ton *gari*) | 832 | 832 | 1205 | 1205 |
| Subtotal | 10156 | 8727 | 13306 | 11229 |
| Interest on capital (8% of subtotal) | 812 | 698 | 1064 | 898 |
| PERFORMANCE MEASURES (Naira) | | | | |
| Total cost per ha | 10968 | 9425 | 14370 | 12127 |
| Cost per ton of *gari* | 3098 | 2662 | 2801 | 2364 |
| Total revenue per ha | 11116 | 11116 | 16108 | 16108 |
| Revenue per ton of *gari* | 3140 | 3140 | 3140 | 3140 |
| Net profit per ha | 148 | 1690 | 1738 | 3981 |
| Net profit per ton of *gari* | 42 | 478 | 339 | 776 |

[a]Firewood is used to toast grated cassava to prepare *gari*.

Note: The exchange rate in 1991 was 17 Naira per US$1.00

and toasting with child rearing. The COSCA team in Nigeria observed several women in Nigeria toasting *gari* and breast-feeding their babies at the same time. The mechanization of any of the harvesting, peeling, and toasting operations will encourage diffusion of the TMS varieties and will encourage farmers who are already planting them to expand the area under cassava cultivation.

## The Delayed Diffusion of the TMS Varieties in Ghana

Why were the TMS varieties released in Ghana only in 1993, some sixteen years after they were released in Nigeria in 1977? This is a puzzle that warrants attention because Ghana is the second-largest producer of cassava in West Africa. Cassava is the cheapest source of calories in Ghana and it competes favorably with maize as the primary food staple, especially in the south. Historically, maize has been the main food staple in Ghana, and the government of Ghana has concentrated its research attention (with the assistance of Centro Internacional de Merjoriento de Maiz y Trigo–CIMMYT) on maize. This explains why the government of Ghana did not display an interest in the TMS varieties when they were introduced in Nigeria in the late 1970s. In fact, Ghana's National Agricultural Research System (NARs) did not initiate a cassava research program until 1980. At that time, cassava breeding was done at the University of Lagon. According to Hahn, however, the university plant breeder was proud of his work and did not encourage the government to import the new TMS varieties.[12] Therefore, Hahn hired a Ghanaian agronomist to help introduce the TMS varieties in Ghana. Hahn reports that the Ghanaian agronomist, Ghana's River Basin Development Authority, funded by the World Bank and Texaco of Nigeria, informally moved truckloads of the planting materials of the TMS varieties from Nigeria to farmers in Ghana during the early 1980s.[13] Yet the government lacked interest in multiplying and distributing the TMS cuttings to farmers.

Around 1984, Ghana's commissioner (minister) for agriculture visited the IITA in Ibadan and met with Hahn. During their discussion, the commissioner used the expression "Monkey de work Baboon de chop" to describe the role of maize and cassava in food policy circles in Ghana. By this he meant that cassava was feeding Ghana but maize was consuming the research resources there."[14] Yet cassava was the crop that survived the severe drought of the early 1980s, and it helped people cope with food insecurity. In 1985, Ghana hosted the Central and Western African Root Crops Network (CEWARN) workshop in Accra. The workshop helped government officials to grasp the importance of cassava in Ghana.

In 1988, eleven years after the TMS varieties were released in Nigeria in 1977, the government of Ghana imported the TMS varieties from IITA and turned them over to Ghanaian researchers for field-testing. Hahn then helped the government of Ghana to obtain IFAD funding for on-farm testing and evaluation of the TMS varieties in Ghana. From 1988 to 1992, the Ghanaian researchers, with the support of an IITA cassava breeder, Dr O. O. Okoli, evaluated the TMS varieties in farmers' fields.[15] The on-farm testing and evaluation revealed that several of the TMS varieties had stable yields across most ecologies in Ghana and had good *gari* and *akple* (pasty cassava food product) qualities. Biochemical tests conducted in 1991 revealed that Ghanaian foods prepared from both the TMS and local varieties had negligible levels of cyanogens. While the field evaluation was going on, entrepreneurs engaged in making cassava starch and dried cassava roots for industrial uses exhibited interest in the TMS varieties (Okoli 2000).

In 1993, sixteen years after the release of the TMS varieties in Nigeria in 1977, the government of Ghana officially released three TMS varieties, which were given local names as follows: TMS 30572, Afisiafi; TMS 50395, Gblemo Duade; and TMS 4(2)1425, Fitaa.[16] In 1998, the TMS varieties were spreading through farmer-to-farmer methods because of the lack of a government distribution program for cassava planting material (Okoli 2000). In 1998, another IFAD-funded project, the Root and Tuber Improvement Program (RTIP) was initiated by the Ministry of Food and Agriculture at Kumasi to multiply and distribute planting cuttings of TMS varieties. In December 2000, the total area planted with the TMS varieties for multiplication in Ghana was 1,142 hectares.

In February 2001, cassava scientists at the Crops Research Institute, Kumasi, reported that the TMS varieties were widely grown by farmers in the Eastern, Greater Accra, and Volta regions, where farmers prepare *gari* for sale in urban areas such as Accra. Farmers in those regions call the TMS varieties "Biafra" (Eastern Nigeria), indicating that the TMS varieties originated from Nigeria. The most common TMS variety in the farmers' fields is Afisiafi (TMS 30572) (Otoo and Afuakwa 2001).

### TMS Diffusion in Uganda

In Uganda, an unusually high incidence and severity of a rare form of the mosaic disease was reported in the Luwero district in 1988. Mosaic-free cuttings of local varieties from relatively unaffected areas within the country were introduced into the severely affected areas. Yet the mosaic disease rapidly spread in the fields planted with the introduced cuttings. Beginning in 1991, TMS 6014, TMS 30337, and TMS 30572 were multiplied, and in 1993 they were distributed to farmers. The multiplication and distribution was financed with assistance from the Gatsby Charitable Foundation of the United Kingdom, United States Agency for International Development (USAID), and some NGOs. By the year 2000, around eighty thousand hectares of the mosaic-resistant TMS varieties were grown in Uganda (Otim-Nape et al. 2000)

### Summary

The TMS varieties are widely diffused in Nigeria and to a lesser extent in Ghana and Uganda. The rapid diffusion of the TMS varieties in Nigeria starting in 1977 was facilitated by the 40 percent increase in farm-level yield and the profitability of the new varieties; the collaboration of the NRCRI, the oil sector, the World Bank, International Fund for Agricultural Development (IFAD), churches, and the Nigerian Cassava Growers' Association; and government oil revenue and the availability of low-cost gasoline. The diffusion in Nigeria was also facilitated by the unflagging leadership of Dr. S. K. Hahn of the IITA. Under Hahn's leadership, the IITA's cassava diffusion program multiplied and distributed TMS planting materials directly to farmers and indirectly through informal channels such as schools and churches. The IITA's diffusion program mobilized the private sector, particularly the oil sector, to assist in the distribution of the TMS varieties. The IITA's cassava program also mobilized the mass media, including the newspapers, radio, and television, to publicize the availability and benefits of the TMS varieties.

The boom in the petroleum industry in the 1970s and the resultant economic expansion in Nigeria led to an increase in the purchasing power of urban consumers and an increase in demand for convenient food products such as *gari*. With the aid of the petroleum revenue, Nigeria was able to improve rural roads; expand agricultural research, extension, and higher education institutions; and finance nationwide demonstrations of the TMS varieties.

Yet it is important to point out that improved cassava food preparation technologies, namely *gari* preparation technologies and the mechanized grater, were in place at the farm-level in Nigeria before the TMS varieties were introduced to farmers in the late 1970s. The availability of mechanized processing equipment and *gari* preparation technologies greatly increased the profitability of the TMS varieties. Finally, farmers were able to utilize mechanized graters, because diesel and gasoline were cheap and readily available.

Mechanized graters have saved labor and they have shifted the labor constraint from grating to the labor-intensive tasks of harvesting, peeling, and toasting. The labor requirements for those tasks increase in direct proportions with yield. The end result is that farmers who have adopted the TMS varieties are often restricted in the amount of cassava they can produce by the labor bottlenecks at the harvesting, peeling, and toasting stages. Mechanization of any of the harvesting, peeling, and toasting tasks will enable farmers to plant a larger area to the TMS varieties.

Ghana's official release of the TMS varieties in 1993, sixteen years after they were released in Nigeria, represents a delayed diffusion, which is attributed to a variety of reasons. First, the government historically invested in maize research, the primary food staple, and initiated cassava research in the NARs only in 1980. Second, some influential Ghanaian scientists did not encourage the government to import the TMS varieties because they were proud of their own breeding work. Third, it took five years, from 1988 to 1993, to evaluate the TMS varieties under field conditions and multiply the stem cuttings. In 1988, Hahn helped the government of Ghana secure financial assistance from international donor agencies to launch a nationwide program of on-farm testing of the TMS varieties.

The diffusion of the TMS varieties in Ghana from 1993 to 2001 has not been documented. In early 2001, however, cassava researchers at the Crops Research Institute, Kumasi, reported that TMS varieties covered large areas of farmers' fields in the Eastern, Greater Accra, and Volta regions, where farmers prepare *gari* for sale in urban centers. The most common of the TMS varieties grown by farmers in those regions is Afisiafi (TMS 30572) (Otoo and Afuakwa 2001). Ghana's diffusion experience can thus be described as a delayed success story.

In Uganda, government and donor interest in the diffusion of the TMS varieties was sparked by the appearance and rapid spread of a rare form of cassava mosaic disease in the late 1980s. Beginning in 1991, TMS 6014, TMS 30337, and TMS 30572 were multiplied. In the year 2000, about eighty thousand hectares of the mosaic-resistant TMS varieties were being grown in Uganda (Otim-Nape et al. 2000).

CHAPTER 8

# The Preparation of Five Cassava Products

## Introduction

Both cassava leaves and roots are prepared as food by different methods in different places in Africa. Simple preparation methods have evolved over time to eliminate cyanogens from cassava roots and leaves in order to make them safe for consumption. These methods are also effective in removing water from cassava roots, which, in turn, extends the shelf-life of cassava and reduces transportation and marketing costs.

There are five common groups of cassava products: fresh roots, dried roots, pasty products, granulated products, and cassava leaves.[1] This chapter examines the evolution of the five cassava products and shows that the traditional methods have been influenced by the availability of water and sunlight, by wage rates, and by market demand in cases where cassava is produced as a cash crop for urban consumption. This chapter also shows that there is regional variation in cassava preparation methods in Africa and that a major challenge is to diffuse the best practices from one region to another within Africa.

### Fresh Cassava Roots

The roots of sweet cassava varieties are eaten raw, roasted in an open fire, or boiled in water or oil.[2] The cyanogens in the roots are destroyed by slowly cooking the roots. Starting with cold water, the roots are gradually heated, promoting the hydrolysis of the cyanogens (Grace 1977).

Boiled cassava roots may be pounded alone or in combination with other starchy staples such as banana (or plantain), yam, cocoyam, or sweet potato. Fresh cassava roots, boiled in water or in oil, are commonly sold by food vendors in major cities and consumed by workers as snacks or for lunch.[3]

Pounded cassava is made from roots boiled in water. The preparation of pounded cassava is elaborate and cumbersome because the boiled cassava roots get sticky during pounding. In Ghana, it takes two women to pound a mortar of boiled cassava roots; one person in a standing position, pounding with a pestle, and the second in a sitting position, removing the sticky dough from the pestle by hand.

### Dried Cassava Roots

Dried cassava roots are stored or marketed as chips, balls, or flour. Chips and balls are milled into flour at home by pounding with a mortar and pestle in preparation for a meal. There are two broad types of dried cassava roots: fermented and unfermented. Sun- or smoke-drying of unfermented dried cassava roots is the simplest method of cassava preparation (fig. 8.1). Smoke-drying requires a large amount of fuel wood. Since this method is inefficient in the elimination of cyanogens, it is used mostly for preparing sweet cassava varieties, which have a low cyanogen content.

In the case of fermented dried cassava roots, the fermentation is accomplished in one of two ways: stacking in heaps or soaking in water. To prepare roots fermented by stacking in heaps, the cassava is cut into pieces, peeled, washed, and left in the sun for a day. The roots are then stacked in a heap, covered with leaves, and left to ferment for about one week. Fermentation in heaps has the advantage of improving the nutritional value of

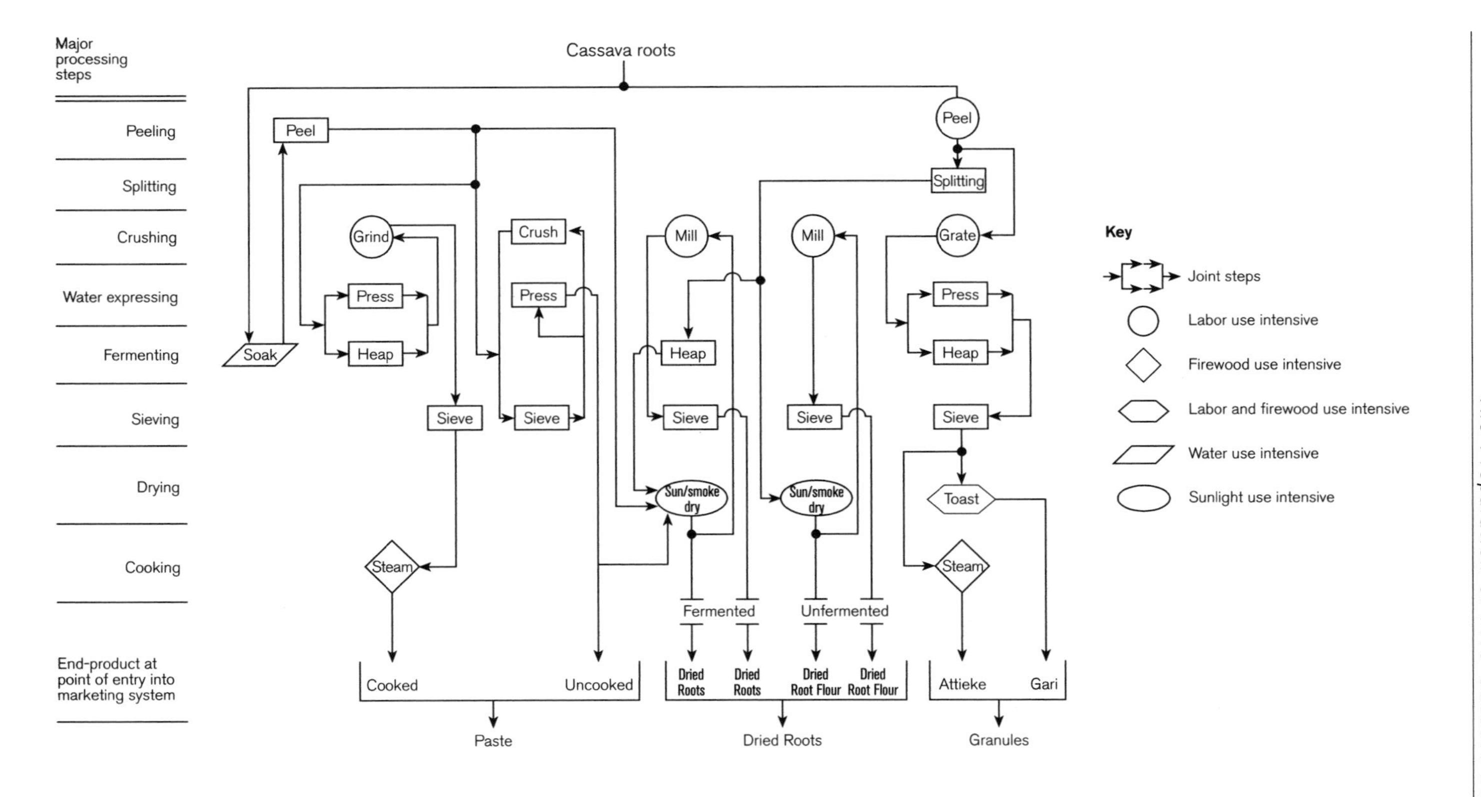


**Figure 8.1.** Traditional Cassava Processing and Food Preparation Methods. *Source: COSCA Study.*

the product. Most of the nutrients that are lost by soaking are retained when fermentation is done by heaping. Additionally, mold growth on the cassava roots in heaps promotes fermentation and increases the protein content of the dried cassava roots (Tanzania Food and Nutrition Cecter and International Child Health Unit n.d.).

The recent introduction of the grater has eliminated stacking and fermentation and therefore has saved time. The roots are now simply peeled, washed, and grated.[4] The pulp is placed in a perforated container and covered, and a weight is put on it for about three hours. The cyanogens are squeezed out along with the effluent. The half-dried pulp is then dried in the sun (Alyanak 1997).

A common fermentation technique is to soak the fresh roots in a puddle or in a pot of water for three to five days. The roots are then peeled (if not peeled prior to soaking) and sun- or smoke-dried directly as whole roots. Alternatively, they can be crushed and pressed to remove the water and molded into balls.

The fermentation process, whether in water or in heaps, influences the taste of the final product. The longer the fermentation period, the stronger the sour taste. Taste is an important attribute, especially for consumers who eat fermented cassava products and who desire a strong sour taste.

The cassava product needs to be visually attractive to a buyer. Bright color is a desired attribute.[5] The brightness of the color of dried roots depends on the method and duration of the fermentation, on the method of drying, on the efficiency of the drying energy, and on the cleanliness of the drying environment.

Cassava can be sun-dried on virtually any surface in the open air, such as a large flat rock in the field, the shoulders of a paved road, flat rooftops, a flat basket, or even bare ground. Although roots dried in a basket may be moved in case of a bad weather, roots being dried on a rock, on the roadside, or on a rooftop often are not easily moved. If it rains, the cassava is soaked, and drying starts all over again. The wide range of drying surfaces used is such that the product can gather significant amounts of mold, dust, and other dirt, which influence the color of the product. In

**Table 8.1.** Type of Cassava Food Products in the Congo, Côte d'Ivoire, Ghana, Nigeria, Tanzania, and Uganda.
*Source: COSCA Study.*

| CASSAVA FOOD PRODUCTS | CONGO | CÔTE D'IVOIRE | GHANA | NIGERIA | TANZANIA | UGANDA |
|---|---|---|---|---|---|---|
| Dried roots | 70 | 8 | 27 | 48 | 91 | 21 |
| Granulated products | 0 | 45 | 43 | 39 | 0 | 0 |
| Pasty products | 25 | 8 | 7 | 13 | 0 | 0 |
| Fresh roots | 5 | 37 | 23 | 0 | 6 | 76 |
| Others | 0 | 2 | 0 | 0 | 3 | 3 |
| TOTAL | 100 | 100 | 100 | 100 | 100 | 100 |

the humid climates, such as in the forest zone, where it rains frequently and where sun-drying is inefficient, the color of the flour is dull white. The savanna zone has a comparative advantage in preparing dried cassava roots because sun-drying is more efficient.

Smoke-drying is done on racks over the kitchen fire. This process imparts a dull color to the product. Processors usually scrape off the dark coating of smoke before it is milled into flour.

The drying process reduces the bulk of the cassava roots, extends shelf life, and thus facilitates cassava marketing in urban areas. Cassava farmers remote from market centers have a comparative advantage in preparing dried cassava roots because dried roots are cheaper to transport to distant urban markets than fresh roots. Dried cassava roots are common in the Congo because of poor roads and limited access to urban markets. When properly dried, cassava roots will keep for several months if they are stored over the kitchen fire; otherwise they may be attacked by weevils or become moldy. Dried cassava roots are common in Tanzania because cassava is used as a famine-reserve crop (table 8.1).

Cassava is eaten in the form of dried roots more often in rural than in urban areas (Idowu 1998). The traditional method of converting dried roots into cassava flour is to pound them in a mortar with a pestle and then sieve the flour through a fine basket or a perforated metal bowl. Consumers can purchase dried cassava roots already milled into flour. Yet cooking the dried root flour is painstaking because it involves stirring the flour in boiling water until it turns into a thick paste.

The level of each resource—i.e., sunlight, fuel wood, water, or labor—needed to prepare dried cassava roots varies with the preparation method used. Sun-drying of cassava roots is sunlight-intensive, while smoke-drying requires fuel wood. Because of the abundance of sunlight, farmers in the savanna usually adopt sun-drying methods, while farmers in the forest zone adopt the smoke-drying method because of limited sunshine. Farmers in the savanna ferment cassava roots by stacking, while farmers in the forest zone ferment by soaking because of the availability of water. Peeling is more labor intensive when it is done prior to rather than after soaking. Milling by pounding with mortar and pestle is always labor intensive.

In most places, women provide the labor used to prepare dried roots. COSCA researchers interviewed a woman in the Congo who was pounding dried cassava roots to make flour for the family dinner. Her husband was sitting under the shed of palm leaves with muddy feet and muddy digging hoe beside him, signs that he was resting after plowing in the field. The COSCA researchers asked the woman why she did not take the dried cassava roots to the nearby mission station to be milled by a machine. Before the woman could respond the man said, "I would not eat cassava flour milled by machine."[6]

### Pasty Cassava Products

Two forms of pasty cassava products are common in Africa. The most popular is called "uncooked paste" because it is stored or marketed without cooking. To prepare the uncooked paste, the roots are soaked in water for three to five days, during which time the roots soften and ferment. The soaked roots are then manually crushed in a fine basket or in a perforated metal bowl in a sack submerged in water. As the soft root is crushed, it is sieved by shaking, thereby separating the pulp into the sack while keeping the fiber in the basket. The sack of pulp is squeezed, often manually or sometimes under a heavy weight, to express the effluent. The end product is a dense pulp, which is stored or marketed.

Preparing uncooked paste is water-use intensive, as additional water is needed, after the soaking, for sieving. Therefore, the preparation of cassava as uncooked paste is concentrated in areas with an abundant water supply. The sieving in water is labor-intensive. In most places women provide the labor to prepare the uncooked paste. This is an unpleasant chore. Women stand in a puddle of water for hours, shaking baskets of the fermented cassava.

Preparing cassava as an uncooked paste extends the shelf life of the cassava product and reduces its volume in comparison with fresh roots. Yet the uncooked paste is not a convenient food product because it needs to be cooked and pounded, sometimes twice, before it is ready for a meal. However, it is commonly used to feed hired labor employed in cassava production. The uncooked paste is less expensive than other cassava products while at the same time it gives a feeling of satiety because it is heavy. In a COSCA study village in southwestern Nigeria, the uncooked paste was nicknamed "six-to-six" by the farmers, meaning 6:00 A.M. to 6:00 P.M. because hired farm labor eat it twice each day, once in the morning and once in the evening, with nothing in between. In some parts of Nigeria, the uncooked cassava paste is transported over long distances in truckloads and retailed in urban markets in small plastic or polypropylene bags.

Cooked cassava pasty products recently have been introduced in Nigerian urban markets. Every evening in major cities in the cassava-growing areas of Nigeria, it is common to find women selling cooked cassava paste wrapped in plastic bags. As women go home from work or from market, they stop and buy some for the evening meal. The paste is best eaten when it is hot. A housewife may buy it in bulk and warm it when it is time to eat. When it is warmed, however, the product forms a jelly on the surface and pounding is required to restore a uniform texture. Although more research is needed on preparation methods, cooked cassava paste is a promising food for busy urban consumers.

Cassava paste can also be stored or marketed in a steamed form. To prepare the paste, fiber is removed by hand from roots fermented by soaking them in water. The roots are then stacked in a heap to further ferment while covered with leaves and pressed with heavy objects to express water. The pulp is ground with a stone or pounded in a mortar. The resulting fine pulp is firmly wrapped in leaves and steamed. There are several variations of this method. For example, in some places the roots are peeled before soaking to improve the attractiveness of the end product. However, peeling before soaking makes the product more expensive because fresh cassava roots require more peeling labor than do roots softened by soaking. In some places, steaming is done twice. After the first steaming the product is kneaded, rewrapped, and then steamed a second time.

Steamed paste is stored or marketed in a ready-to-eat form. Preparing steamed paste is expensive because many steps are involved and each one requires additional inputs. For example, grinding and sieving are labor-intensive. The soaking step is water-use intensive, and steaming is fuel wood-use intensive. In the Congo, steamed paste, *chickwangue,* is prepared entirely by women. It is also common throughout Central Africa, from Cameroon to Burundi (Jones 1959).[7] Steamed paste is not as bulky as fresh cassava roots and therefore it is less expensive to transport. The double fermentation as well as the steaming impart a long shelf life to steamed paste. The sour flavor achieved through extended fermentation is a characteristic that is cherished by regular customers. Yet it is also a turnoff to potential new consumers.

The color of the pasty products (uncooked or steamed) depends upon the quality of the water in which the cassava is soaked. Cassava may be soaked in running water, such as a stream or a river, or in stagnant water, such as a ground puddle or water in a container, depending on what is available. If cassava is soaked in clean running water, the pasty product will have a bright white color. If the soaking is done in dirty water or over an extended period, the product acquires a dull color. In the Congo, the regions of Bandundu and Bas-Congo are suppliers of steamed paste to the urban population of Kinshasa, where consumers prefer Bas-Congo products that are brighter in color. The brownish tint in cassava paste in the Bandundu region has been traced to the color of the water in the streams in which the cassava is soaked (Osiname et al. 1988).

### Granulated Cassava Products

There are three common types of granulated cassava products: *gari, attieke,* and *tapioca.* The methods for making granulated cassava products originated in Brazil. As we pointed out in chapter 1, the diffusion of cassava from South America to West Africa was delayed until the methods of preparing cassava as granulated products were also introduced from South America.

*Gari* is a toasted food product that is much more common in Nigeria and Ghana than in the other four COSCA study countries. To prepare *gari,* fresh cassava roots are peeled, washed, and grated. The resulting pulp is put in a porous sack and weighted down with a heavy object for three to four days to express effluent from the pulp while it is fermenting. The dewatered and fermented lump of pulp is pulverized and sieved and the resulting semi-dry fine pulp is toasted in a pan. Palm oil is sometimes added during toasting in order to prevent the pulp from burning. However, the addition of palm oil changes the color of the *gari* from white to yellow. The grating, effluent expression, pulverization, toasting, and addition of palm oil are adequate to reduce cyanogens to a safe level (Hahn 1989).

Fermentation imparts a sour taste to *gari.* The duration of fermentation varies depending on consumer preference for sour taste. The COSCA study found that commercial *gari* processors in Nigeria ferment cassava for different lengths of time, depending on the market. For example, in Edo State, a major commercial cassava-producing state, *gari* preparers ferment for a maximum of three days for *gari* prepared for markets in Eastern Nigeria and for five days or more for markets in Western Nigeria. The strong sour taste in *gari* acquired from extended fermentation is appreciated by the Yoruba people of Western Nigeria. The Ibo people of Eastern Nigeria prefer *gari* that is bland in taste. Toasting extends the shelf life so that *gari* easily can be transported to urban markets. If kept in a dry environment, *gari* will store better than grain because *gari* is not known to be attacked by weevils (Okigbo 1980).

*Gari* preparation tasks are labor-intensive, and women provide the manual labor for these tasks. However, in the next chapter we shall discuss a mechanized method of cassava grating that is widely used in Nigeria and Ghana. A large amount of fuel wood is needed for the toasting. Many buy fuel wood and hire female labor to prepare *gari* for sale to urban consumers.

To summarize, *gari* is a convenient product because it is stored and marketed in a ready-to-eat form. It can be soaked in hot or cold water, depending upon the type of meal being prepared. *Gari* has a long shelf life, and it is therefore attractive to urban consumers. It is the most common form in which cassava is sold in Ghana (Doku 1969) and in Nigeria (Ngoddy 1977).

The second type of granulated cassava product is *attieke,* a type of steamed cassava that is found only in the Côte d'Ivoire. *Attieke* is made in much the same way as *gari,* with more or less the same inputs. Instead of toasting, however, *attieke* is steamed. *Attieke* is available in a wet form, and it has a shorter shelf life than does *gari.*

The third type of granulated cassava product is *tapioca,* which is primarily consumed in Togo and Benin.[8] To prepare *tapioca,* cassava is grated and then put in water, pressed, and kneaded to release the starch. The starch is permitted to settle at the bottom of the container, and the water is drained off. The operation is repeated several times to prepare high-quality product. The damp starch is spread in a pan and toasted in the same way that *gari* is, to form a coarse granular product. It may later be sifted to sort the granules by size.

## Cassava Leaves

Cassava leaves are edible and a more convenient food product than are fresh roots. Cassava leaves have a nutritive value similar to that of other dark green leaves and are an extremely valuable source of vitamins A (carotene) and C, iron, calcium, and protein (Latham 1979). The consumption of cassava leaves helps many Africans compensate for the lack of protein and some vitamins and minerals in the roots. Cassava leaves are prepared by leaching them in hot water, pounding them into pulp with a mortar and pestle, and then boiling them in water along with groundnuts, fish, and oil. This process eliminates cyanogens from the leaves, making them safe for human consumption.

Cassava leaves are an important vegetable in the Congo and in Tanzania. The COSCA study found that farmers in the Congo select cassava varieties with a large canopy of leaves. In Tanzania, farmers plant "tree cassava," *mpiru,* for the production of leaves. In countries where cassava leaves are eaten as vegetables, producers earn additional income by selling cassava leaves. Truckloads of cassava leaves, locally called *pondu* in the Congo, are a common sight on the roads from the provinces to Kinshasa.

Cassava leaves are not eaten in Uganda because they are held in low regard. Their consumption is considered an indicator of low economic status (Otim-Nape 1995). Cassava leaves are not eaten in West Africa, except in Sierra Leone, because several indigenous plants supply vegetables traditionally consumed with yam (Okigbo 1980). Most of these vegetables are, however, available only during the rainy season. Therefore, there is a seasonal gap in the availability of vegetables in West Africa which cassava leaves could fill. For example, an assessment of the nutritional status of children in parts of the forest zone of southwest Nigeria shows that between February and May, many children are malnourished, and *kwashiorkor* cases are common and infant mortality is high (Nnanyelugo et al. 1992). The period from February to May is a dry period, and green vegetables are scarce, but cassava leaves are available because cassava grows leaves all year round.

Women in the cassava-growing areas of West Africa are generally not aware that cassava leaves are edible and rich in protein, vitamins, and minerals. The women do not know that their children can be saved from malnutrition and even death by feeding them cassava leaves in the dry months when other vegetables are in short supply. The consumption of cassava leaves as a vegetable will make cassava production more profitable and increase the food security and nutritional status of African families.

Cassava leaves can be stored in dry form, and since they have a lower water content, they are less expensive to dry than the roots. If leaf harvesting is properly scheduled, it does not have an adverse effect on cassava root yield. For example, when leaves were harvested monthly, the reduction in root yield ranged from 22 to 42 percent (Dahniya 1983). Root yields were not affected, however, when the leaves were harvested every two months (Lutaladio and Ezumah n.d.).

## Lessons and Challenges

A lesson that emerges from this chapter is that the traditional methods used to make five cassava products have been influenced by the availability of sunlight, water, and fuel wood and by wage rates. For example, in areas where water is scarce, the sun-drying method is popular. In areas where sun-drying is not efficient, smoke-drying is practiced. Likewise, farmers who produce cassava as a cash crop process it as *gari* by purchasing fuel wood and hiring female labor. This diversity of natural resources and differentiation in wage rates has contributed to an array of cassava products in Africa.

Another important lesson that emerges from the COSCA study is the localized nature of cassava preparation methods. For example, *gari* preparation is common throughout West Africa but not in Central and East Africa. The consumption of cassava leaves is common in the Congo and Tanzania but not in the Côte d'Ivoire, Ghana, Nigeria, or Uganda. The challenge is to promote the diffusion of the best practices, such as the preparation of *gari* and cassava leaves, throughout the cassava-growing areas of Africa.

Nigeria and Tanzania provide an example of the mutual benefits of exchanging information on best practices. The Nigerian method of *gari* preparation could be introduced in Tanzania. Likewise, the Tanzanian best practices in the preparation of cassava leaves could be introduced in Nigeria, where cassava leaves are not now consumed. In 1994, under the aegis of the COSCA study, two Tanzanian homemakers were brought to Nigeria and paired with *gari* preparation host families to understudy how to make *gari*. The activity was interrupted by the civil strife that followed the annulment of the 1993 presidential election in Nigeria. However, we learned subsequently that people in the villages of the two Tanzanian homemakers were mixing *gari* with maize to make *ugali,* the most common traditional maize meal in Eastern and Southern Africa (Kapinga 1995).

In summary, diversity of natural resources and differentiation in usage rates have contributed to an array of cassava products in Africa. The challenge is to promote the best practices, such as the preparation of *gari* and cassava leaves, throughout the cassava-growing areas of Africa.

*Bundles of cassava stems for use as planting material for sale to farmers.*

*A field of the TMS cassava varieties. The leaf density reduces weed growth.*

*Nigerian farmers harvesting cassava.*

*Fresh cassava roots for sale in a market.*

*Women and children peeling cassava with sharp knives in readiness for grating using mechanized graters.*

*Mechanized grating of cassava for a female farmer in a Nigerian village.*

*A Ghanaian woman preparing* gari *by toasting grated cassava root. In Africa, women often combine cassava processing with childcare.*

Gari *for sale along with yam (background) in Nigeria.*

## CHAPTER 9

# Mechanized Processing

## Introduction

We pointed out in chapters 5 and 6 that farmers in Nigeria who planted the TMS varieties had difficulties in recruiting labor for harvesting and manual processing. In order to eliminate hand pounding and reduce processing costs, research has been carried out in Nigeria and several other countries to develop mechanized processing machines. In this chapter we shall trace the evolution of public and private attempts to develop improved cassava processing machines. We shall show that private sector smiths, welders, and mechanics have been more successful in developing cost-effective machines than have researchers in government research institutes.

## Types of Cassava Processing Machines

There are three common types of cassava processing machines in use in Africa: graters, pressers, and mills. The graters are mechanized grating machines that convert fresh cassava roots into mash for preparing dried cassava roots and granulated products. The pressers are used to press effluent from cassava mash. The mill is used for converting any food crop in dry form, including dried cassava roots, into flour.

### ■ The Mechanized Grater

Traditionally, cassava was pounded in a mortar with a pestle. Later, artisans developed a manual grater in the form of a sheet of perforated metal mounted on a flat piece of wood. The cost of the manual grater was modest because it was made from a piece of wood, scrap metal, and a few nails. The hand grater was easy to wash and dry. However, the efficiency of the hand grater was low and women who used it sometimes sustained hand injuries.

Jones (1959, 209) reports that the French introduced mechanical graters in Benin (formally Dahomey) in the 1930s to teach farmers how to prepare *gari* and *tapioca* for export markets. Adegboye and Akinwumi (1990) report that a hand-operated mechanical grater was developed in 1931 by an artisan in Oyo, a town in southwestern Nigeria. The artisan placed one of the machines on the roadside in Oyo, where some expatriate staff of the United Africa Company (UAC) noticed it and helped the artisan develop the first engine-driven cassava grater in Nigeria (Adegboye and Akinwumi 1990).[1]

The COSCA study found that mechanized graters in villages in Nigeria and Ghana have been developed and refined over time by village smiths, welders, and mechanics. The mechanized graters are made with old engines and scrap metals. The cost of a grater ranges from US$200 to US$500. Most of the graters are owned by village entrepreneurs and operated by young men or teenage boys who provide grating services to smallholders for a fee based on the quantity grated. The quantity processed for a customer can be as small as one kilogram or as large as several tons. The processors are at the beck and call of farmers at any hour of the day. In some villages, the graters are located in a marketplace. In other villages, a grater is mounted on wheels and wheeled to the fields or homes of farmers who request its services.

In many villages, local machine operators provide a comprehensive set of services, including mechanized grating, pressing, and sieving of the

**Table 9.1.** Percentage of Villages in COSCA Countries with Mechanized Cassava Processing Machines.
*Source: COSCA Study.*

| MACHINE | CONGO | CÔTE D'IVOIRE | GHANA | NIGERIA | TANZANIA | UGANDA |
|---|---|---|---|---|---|---|
| Grater | 0 | 10 | 17 | 52 | 0 | 0 |
| Presser | 0 | 7 | 0 | 30 | 0 | 0 |
| Miller | 1 | 18 | 83 | 55 | 33 | 62 |

cassava. The components of the mechanized grater can be changed so that the same engine can be used both to pulverize and to sieve the cassava mash. In more comprehensive village processing centers, farmers also toast *gari* at the processing center. Maintenance services for the graters are provided by roadside mechanics and welders at any hour of the day.

In Nigeria and Ghana, the mechanized grater is now being used by some commercial *gari* processors (table 9.1). The replacement of hand graters with the mechanized grater has reduced the cost of producing *gari*. The COSCA study found that in Nigeria, fifty-one days of labor were needed to prepare a ton of *gari* by hand, while only twenty-four days were required to prepare the same amount by mechanized grater. Table 9.2 compares the cost of production per ton of *gari* with TMS varieties and

**Table 9.2.** Nigeria: Cost per Ton of *Gari* with TMS Varieties and Manual versus Mechanized Grating, 1991.
*Source: COSCA Study.*

| YIELD (TONS/HA) | MANUAL GRATER | MECHANIZED GRATER |
|---|---|---|
| 3 | 43,993 | 39,247 |
| 6 | 28,693 | 23,447 |
| 9 | 23,233 | 18,747 |
| 12 | 21,043 | 16,196 |
| 15 | 19,514 | 14,664 |
| 18 | 18,493 | 13,647 |
| 21 | 17,763 | 12,917 |
| 24 | 17,217 | 12,367 |
| 27 | 16,793 | 11,947 |

Note: 85 Naira = US$1.00

manual versus mechanized grater. For example, when the average cassava yield is 18 tons per hectare, the total cost of making a ton of *gari* with manual grating is 18,493 Naira, as compared with 13,647 Naira per ton with mechanized grating.

To summarize, the mechanized grater has reduced processing labor by 50 percent and dramatically increased the profitability of *gari* production in Nigeria and Ghana. Since the grating task has been mechanized, peeling is now the most labor-intensive task in *gari* processing, followed by the toasting stage.

COSCA researchers discovered that the mechanized grater was infrequently used by *attieke* processors in the Côte d'Ivoire because the texture of *attieke* made with the mechanized grater was not as good as that of hand-made *attieke*. The planting of the TMS varieties is thus not profitable to farmers who process cassava as *attieke* because manual grating is labor-intensive.

## ■ The Mechanized Presser

Since cassava has a high (70 percent) water content, various methods have been developed to extract the water during processing. Traditionally, effluent was drained from grated cassava pulp through a process by which it was put in a basket, covered with leaves, and pressed with a heavy object such as a stone for three to five days. Fermentation took place at the same time. However, pressing has evolved through three stages. First, instead of a basket, the mash was placed in a sack. Second, the stone was replaced with wooden plates. The sack of cassava pulp was tightened between two wooden plates with a rope to squeeze out the effluent. The time period taken to express the effluent was thus reduced from around four days to two and the period of fermentation was also cut from four to two days.

The most recent development was the replacement of the rope with a screw press.[2] Using this method, a sack of cassava mash is placed between two wooden plates and pressed with an automobile jack to remove the effluent (Adegboye and Akinwumi 1990). The COSCA study found that the mechanized presser was introduced as early as 1940 in the Côte d'Ivoire but it was only in the 1960s that it was widely adopted (fig. 9.1). The screw

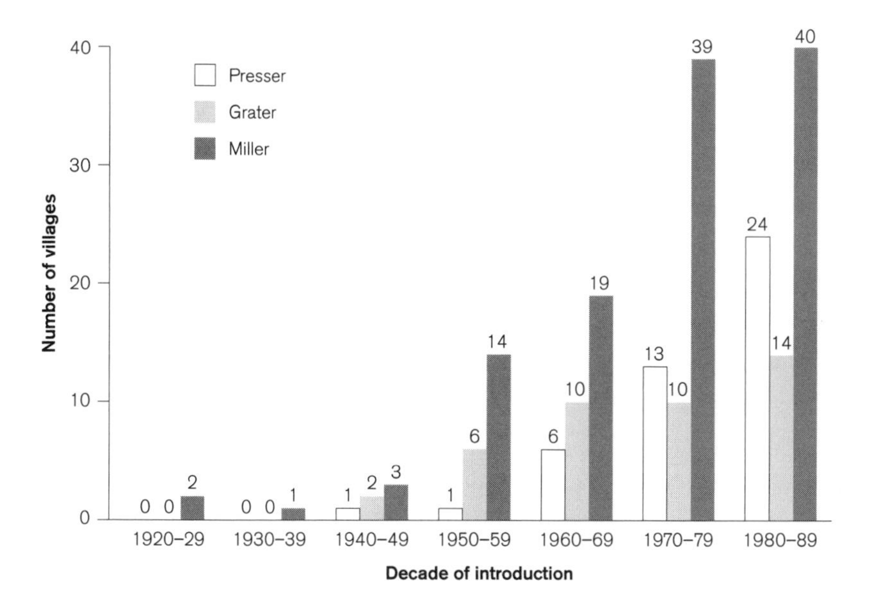


**Figure 9.1.** History of Introduction of Cassava-Processing Machines in Villages in the Six COSCA Countries.

*Source: COSCA Study.*

press is currently used by many commercial *gari* processors in Nigeria and most *attieke* processors in the Côte d'Ivoire. The screw press eliminates the fermentation step and reduces the processing time from several days to a few hours.

The mechanized presser is a simple hand-operated machine constructed from local wooden plates and a used automobile jack. Mechanized pressers are operated and maintained by village entrepreneurs. In Nigeria, mechanized graters and pressers are often operated and maintained by the same village entrepreneurs in the major areas of commercial cassava production. This enables farmers to have access to grating, pressing, pulverizing, and sieving services in one convenient location.

The advantage of the presser is that it removes the effluent in a few hours, but the inherent disadvantage is that it eliminates the fermentation step. However, fermentation is not essential for the elimination of cyanogens in the preparation of *gari* because the grating, pressing, pulverizing, sieving, and toasting steps reduce the cyanogen content in cassava to such a low level that *gari* is safe for consumption. Yet, fermentation imparts a sour taste, which is cherished by some consumers in southwestern Nigeria. Taste is a serious consideration in consumer acceptance of a new cassava product.

### ■ Mechanized Food Crop Mills

The traditional method of preparing cassava flour from dried cassava roots is to pound the roots in a mortar with a pestle. Mechanized milling machines were introduced in Nigeria in the1920s but they did not gain widespread adoption until the 1950s. Mechanized mills are used to convert dried food crops, including dried cassava roots, dried yam tubers, grains, beans, and peas into flour. The components of the mechanized mill are also fabricated locally from scrap materials. The mill is usually owned, operated, and maintained by village entrepreneurs. The mill is generally located in a village center such as a marketplace. Farmers, processors, and homemakers from the village take their food crops to be milled for a small fee per unit of quantity.

Today, mechanized mills are commonplace in the urban centers in the six COSCA study countries. Mills were also observed in several COSCA villages in Nigeria, Ghana, and Uganda. However, mills were found in only a small percentage of the COSCA villages in the Côte d'Ivoire, Tanzania, and the Congo. During the COSCA study in the Bandundu region of the Congo, the sound of pounding reverberated in villages from late afternoon into the night as women pounded dried cassava root (*cossette*) to make flour for dinner. This use of labor can be described as an onerous waste of effort.

## The Rise and Fall of Private Large-Scale Processing

In 1931, a private Ghanaian (formerly Gold Coast) entrepreneur designed and patented an automated grater and screw press.[3] He applied to the government for a loan to finance the construction of a large-scale plant to process cassava. The plant was closed in 1958 because it was unprofitable (Doku 1969).

In the 1970s, the fast-growing urban population and increasing demand for food spurred the development of several industrial schemes for processing and preparation of traditional cassava foods. These schemes included the Texagri and Root Crop Production Company in Nigeria, Sodepalm and I2T in the Côte d'Ivoire, Socamrico and Siac in Cameroon, and Mantsoumba State Farm and Agricongo in the Congo. Many of these companies have now ceased operation because they were unprofitable owing to an irregular supply of cassava roots and an inability to prepare cassava products to meet the color, taste, and texture requirements of different ethnic populations (Bokanga 1992).

Many large-scale processing schemes failed also because of technical and managerial problems. The first problem faced by such schemes is the difficulty of hiring and managing a large number of women to hand-peel cassava. The production capacity of a plant producing from 2 to 8 tons of dry cassava products per day requires a daily input of twenty-two to eighty-eight woman-days of peeling labor. A second technical problem is related to fermentation. Large-scale industrial operators have difficulty producing fermented cassava with a uniform taste. Successful cassava processing plants must have skilled technicians with a good understanding of cassava fermentation and its potential for detoxifying cyanide, protecting against microbial contaminants, and producing desirable flavors in cassava products. A third technical problem is the variability in shape and size of cassava roots. This variability must be addressed by hand-peeling because the irregularities are too great for mechanization to be effective. The development of cassava varieties that produce roots with uniform shape and size is thus a major challenge for cassava breeders and processing engineers.

### Nigeria: Cassava Processing Research and Development

In 1955, the Nigerian colonial government established the Institute of Applied Industrial Research to undertake research into the chemical and biochemical makeup and engineering and processing of crops, including cassava (Idachaba 1998).[4] The institute was redesignated as the Federal Institute of Technical Research in 1958, and the name was changed to the Federal Institute of Industrial Research, Oshodi (FIIRO) in 1975. The FIIRO designed a mechanized processor consisting of a peeler, washer, grater, sieve, and toaster all in one integrated machine. The processor was fabricated in Britain and returned to Nigeria but it was not useful to the smallholders and small-scale processors, who needed flexibility in the quantity of cassava handled, working hours, and proximity to the villages.

In 1971, two new government agencies, the Products Development Agency (PRODA) and the Fabrication Engineering and Production Company (FABRICO), were established to develop cassava-processing machinery using local material. The agencies developed intermediate-sized graters. In 1981, the Rural Agro Industrial Development Scheme (RAIDS) was established by the federal government to test the PRODA- and FABRICO-designed machines and to identify farmers' food crop processing problems (Idowu 1998). In 1983, the African Regional Centre for Engineering Designs and Manufacturing (ARCEDEM) based in Ibadan became the fourth public-sector agency to be engaged in cassava postharvest engineering research and development activities. The center was charged with verifying designs and fine-tuning prototype machines in collaboration with RAIDS. Private engineering companies were contracted by the RAIDS agency to fabricate prototypes for distribution to cassava processors (Idowu 1998).

The graters developed by the government agencies have achieved limited adoption because they are more expensive and not as efficient, reliable, or convenient as those developed by the village artisans. Also, the graters developed by engineers in the government agencies have capacities far in excess of the processing needs of the smallholders. The entrepreneurs who bought the government machines have either had them modified by local artisans or abandoned them (Adegboye and Akinwumi 1990).

### Lessons and Challenges

There are three common types of cassava-processing machines in use in Africa: graters, pressers, and mills. The traditional method of grating cassava was by pounding it in a mortar with a pestle. Later, artisans developed a manual grater in the form of a sheet of perforated metal mounted on a flat piece of wood. Mechanized graters were first introduced in the Benin Republic by the French in the 1930s and later modified in Nigeria in the 1940s by welders and mechanics, using local materials such as old automobile motors and scrap metal. The mechanized graters are owned by village entrepreneurs, who provide grating service to farmers. The mechanized grater operators allow the farmers flexibility in terms of working time and quantity of cassava grated. The fee charged is a small fraction of the cost of grating by hand.

In Nigeria and Ghana, the mechanized grater is used by commercial *gari* processors. The use of a mechanized grater has reduced *gari* processing costs by 50 percent and has dramatically increased the profitability of *gari* production with the TMS varieties in these countries. Since the grating task has been mechanized, peeling is now the most labor-intensive task in *gari* preparation, followed by toasting. In the Côte d'Ivoire, the mechanized grater is not used by many *attieke* processors because the texture of *attieke* made with the mechanized grater is not as good as that of hand-made *attieke.* The planting of the TMS varieties is thus not profitable to farmers who process cassava as *attieke,* because hand-grating is labor-intensive.

A variety of methods have been developed to extract the water from cassava during processing. In the traditional method, effluent is drained from grated cassava mash by putting it in a basket, covering it with leaves, and placing a heavy object such as stone on top of it for three to five days. Fermentation takes place at the same time. However, most commercial *gari* makers in Nigeria and *attieke* makers in the Côte d'Ivoire now use a screw jack to extract the effluent. The mechanized presser is a simple hand-operated machine, which is made from wooden plates and a used automobile jack, both of which are available in villages. In Nigeria, mechanized graters and pressers are often operated and maintained by the same village entrepreneurs in major areas of commercial cassava production. This enables farmers to have access to grating, pressing, pulverizing, and sieving services all in one convenient location.

Mechanized pressing reduces the processing time from several days to a few hours, yet the inherent disadvantage of the mechanized presser is that it performs only the pressing step and eliminates the fermentation step. Fermentation imparts a sour taste which is cherished by consumers in some places, such as in southwestern Nigeria. However, fermentation is not essential for the elimination of cyanogens in the preparation of *gari* because the grating, pressing, pulverizing, sieving, and toasting steps reduce the cyanogen in cassava to such a low level that *gari* is safe for human consumption.

The traditional method of preparing cassava flour from dried cassava roots is to pound the roots in a mortar with a pestle. Today, mechanized mills are commonplace in the urban centers in the six COSCA study countries. Mills were also observed in several COSCA villages in Nigeria, Ghana, and Uganda, though in only a small percentage of the COSCA villages in the Côte d'Ivoire, Tanzania, and the Congo. The components of the mechanized mill are also fabricated locally from scrap materials.

The following new challenges in mechanized cassava processing flow from the COSCA study:

- The mechanized grater and presser in use in Nigeria and Ghana should be introduced in the other major cassava-producing countries in Africa. The grater can be made from local materials, and the machines are reliable and cost-effective.
- Engineers and breeders should join forces to mechanize the peeling task by developing cassava roots that are uniform in shape and size and have skins that meet certain specifications.
- New technology is needed to reduce the amount of labor required for the toasting stage of *gari* preparation.
- A mechanized grater should be developed that will make attieke with a texture as good as that of hand-made *attieke*.

In the past, attempts have been made in the various COSCA countries by private entrepreneurs to process cassava on a large scale using integrated grating, pressing, pulverizing, sieving, and toasting machines in a factory style of operation. However, these firms failed because of an array of problems, including shortage of cassava roots, fragmented market for cassava products, unacceptable consumer taste, and high cost of labor for hand-peeling.

Numerous public-sector agencies were established in Nigeria in the 1970s and 1980s to develop and diffuse mechanized cassava-processing machines, yet the machines developed by these government agencies have not been adopted by farmers because the machines are not as convenient or reliable as those developed by the small-scale private artisans.

Important lessons for mechanization of food processing in Africa emerge from this analysis of the evolution of mechanized cassava-processing methods. Mechanization of food processing must be done only for products that have a market demand, such as *gari.* Mechanized machines should be small and easy to fabricate and repair by village artisans using local materials, such as old engines and scrap metals. Since most villages in Africa do not have electricity, and petroleum fuel is scarce and expensive, machines that are manually driven are more suitable for remote villages, provided they can be fabricated and maintained by village artisans using local materials.

CHAPTER 10

# Gender Surprises

## Introduction

The gender debate mushroomed in Africa in the 1970s following the publication of Esther Boserup's influential *Women's Role in Economic Development* (1970). Boserup, a Danish social scientist, provided an array of evidence to show that women in developing countries play significant roles in agricultural and rural development and that Africa was the region of female farming par excellence. She reported that in many African tribes, nearly all the tasks connected with food production are carried out by women.[1] Boserup drew on eighteen anthropological village studies and concluded that women in Africa often "do more than half of the agricultural work; in some cases . . . around 70 percent, and in one case nearly 80 percent of the total" (22).

Based on the evidence from anthropological case studies and surveys, Boserup urged researchers, policy makers, and donor agencies to give more attention to the role of women in research and in development projects. She argued that "women usually lose in the development process," because men monopolize the use of new equipment and agricultural methods, and this tendency is frequently reinforced by a bias in extension programs in favor of men. As a result, there will be a relative decline in the productivity of women, and "the corollary of the relative decline in women's labor productivity is a decline in their relative status" (Boserup 1970, 53).

Inspired but puzzled by Boserup's sweeping claim that "women lose in the development process," Dunstan Spencer (1976) carried out a study of a new rice production project in the Sierra Leone and collected biweekly data on the workload of men, women, and children engaged in rice production. Spencer and members of his research team interviewed twenty-three rice-producing families in the Sierra Leone once every two weeks from May 1974 to June 1975. A total of fourteen of the twenty-three households were participating in a World Bank–financed project to increase rice production in the eastern province of Sierra Leone, while the other nine households were nonparticipants selected at random from the same province. Spencer's meticulous study of the numbers of hours worked by each member of the twenty-three households revealed that the workload of the women in the households participating in the rice project increased only slightly, while the workload of the men and male children was substantially increased during the first three years of the rice development project.

Based on these findings, Spencer rejected Boserup's hypothesis that women's workload increases relative to men's as commercialization of agriculture proceeds. However, Spencer acknowledged that his sample was small and that more research was needed on the impact of technical change on labor inputs and the returns to the labor of men, women, and children in different ethnic groups and in different rice production systems. In the final analysis, the critical policy question is the returns to an hour of labor worked rather than number of hours worked. Unfortunately, there are very few rigorous empirical studies in the literature of the workload and returns per hour of labor by gender over time.

Without question, women have played and will continue to play significant roles in food production in Africa. Yet the literature inspired by Boserup's book has generated some sweeping generalizations, such as that women produce up to 80 percent of the food in Africa. Virtually every major donor agency and many gender and agriculture experts have reported that women account for 70 to 80 percent of African food production (Snyder 1990; Brown et al. 1995; Food and Agriculture Organization of the United Nations 1997; USAID 1997; World Bank 2000). Also, a number of scholars have reported that women were denied access to resources needed to produce the food, such as land, credit, and extension advice (Udry et al. 1995; Quisumbing 1996; Lahai, Goldey, and Jones 2000). Gladwin recently reported that the Boserup thesis is by now "the accepted wisdom" that women are losers in the development process (1996, 127).

COSCA studies of the cassava food system in six countries in Africa have generated several gender "surprises." We shall synthesize COSCA evidence and show that *both* men and women allocate significant amounts of their labor to cassava production in countries where cassava is a major food crop. We shall also show that men and women typically perform different and changing roles during the cassava transformation. Yet it is unwise to generalize for all of Africa.

## Cassava Production

In each of the six study countries, COSCA researchers collected information on male and female labor inputs by cassava fields and by tasks such as land clearing, plowing, planting, weeding, harvesting, transporting, and processing. Figure 10.1 reveals that the number of cassava fields in which women provided the bulk of the labor for each task increased from a low level during land clearing to a higher level at weeding, harvesting, transporting, and processing.[2] By contrast, the number of fields in which men provided the bulk of the labor was highest during land clearing, plowing, and planting. These findings show that *both* men and women are heavily engaged in different cassava production tasks in the six most important cassava-producing countries in Africa (table 10.1). To summarize, men typically work on land clearing, plowing, and planting while women perform weeding, harvesting, transporting, and processing tasks.

The division of labor by task is influenced by many factors, including the power intensity of the different tasks that are performed (Spencer

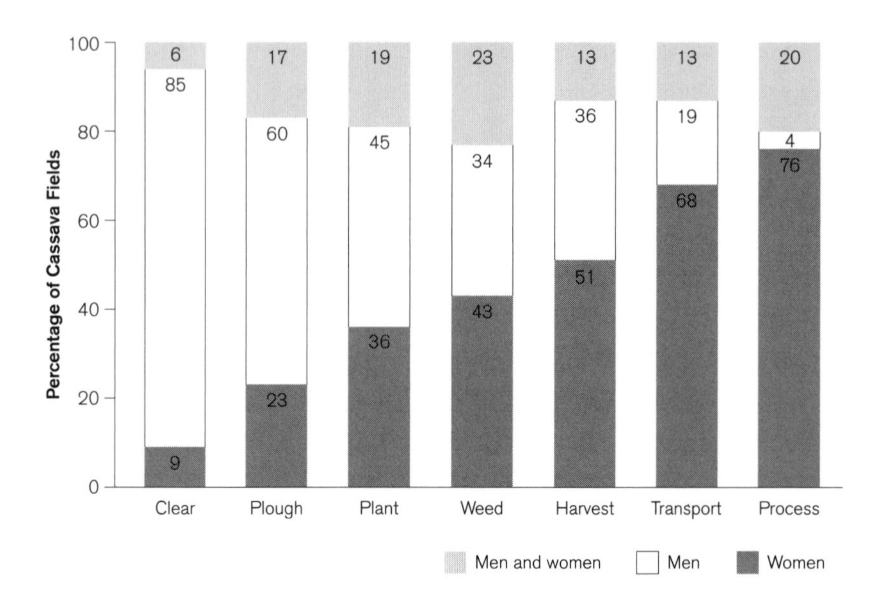


**Figure 10.1. Percentage of Cassava Fields in which Men, Women, or Men and Women together, Provided the Bulk of Labor for Different Tasks, Average for Six COSCA-Study Countries.** ***Source: COSCA Study.***

1976). For example, manual land clearing and plowing for cassava planting are power-intensive tasks that are usually performed by men. In some isolated cases, however, women also perform some of these power-intensive tasks. A COSCA researcher in the Congo asked a male farmer who provided the bulk of the labor for transporting cassava from his field. The farmer retorted that the COSCA investigator should know that only women could perform this task because transporting cassava by head- or back-load is a very heavy task.

The Congo is a textbook example of a country at the rural food staple stage of the cassava transformation. In areas of the Congo where there is little or no tree crop production, women do virtually all the cassava work because young men have outmigrated in large numbers to try to find employment in urban centers (Drachoussoff, Focan, and Hecq 1993). In the COSCA study villages, women in households with low ratios of men

**Table 10.1.** Percentage of Household Cassava Fields in which Women and Men Provided the Bulk of Labor for Field Tasks in Six COSCA-Study Countries. *Source: COSCA Study.*

| COUNTRY | LAND CLEARING | PLOUGHING | PLANTING | WEEDING | HARVESTING | TRANSPORTING | AVERAGE |
|---|---|---|---|---|---|---|---|
| Congo | | | | | | | |
| Women | 27 | 75 | 85 | 84 | 85 | 92 | 82 |
| Men | 68 | 9 | 5 | 6 | 2 | 1 | 17 |
| Both equally | 5 | 16 | 10 | 10 | 13 | 7 | 1 |
| Total | 100 | 100 | 100 | 100 | 100 | 100 | 100 |
| Côte d'Ivoire | | | | | | | |
| Women | 3 | 8 | 38 | 51 | 31 | 79 | 41 |
| Men | 90 | 90 | 46 | 31 | 40 | 8 | 54 |
| Both equally | 7 | 2 | 16 | 18 | 29 | 13 | 5 |
| Total | 100 | 100 | 100 | 100 | 100 | 100 | 100 |
| Ghana | | | | | | | |
| Women | 1 | 1 | 25 | 10 | 24 | 78 | 21 |
| Men | 97 | 94 | 52 | 70 | 48 | 1 | 63 |
| Both equally | 2 | 5 | 23 | 10 | 28 | 21 | 16 |
| Total | 100 | 100 | 100 | 100 | 100 | 100 | 100 |
| Nigeria | | | | | | | |
| Women | 4 | 4 | 24 | 34 | 30 | 83 | 26 |
| Men | 94 | 95 | 71 | 62 | 66 | 10 | 72 |
| Both equally | 2 | 1 | 5 | 4 | 4 | 7 | 2 |
| Total | 100 | 100 | 100 | 100 | 100 | 100 | 100 |
| Tanzania | | | | | | | |
| Women | 8 | 12 | 22 | 25 | 43 | 17 | 21 |
| Men | 77 | 51 | 39 | 19 | 22 | 51 | 42 |
| Both equally | 15 | 57 | 39 | 56 | 35 | 32 | 37 |
| Total | 100 | 100 | 100 | 100 | 100 | 100 | 100 |
| Uganda | | | | | | | |
| Women | 9 | 13 | 22 | 33 | 53 | 74 | 36 |
| Men | 73 | 55 | 41 | 32 | 25 | 15 | 42 |
| Both equally | 18 | 36 | 37 | 35 | 22 | 11 | 22 |
| Total | 100 | 100 | 100 | 100 | 100 | 100 | 100 |

to women carried out a substantial proportion of land clearing and a significantly larger proportion of plowing for cassava production than did men.[3] Yet the reverse pattern was found in the tree crop-dominated rural economies in the Congo where men have remained in the villages because they can earn a higher wage in tree crop production than in the urban centers. The COSCA study found that the proportion of cassava work done by men was higher in tree crop- than in non-tree crop-producing households.

In the Congo, women are left to produce food for their families under conditions of extreme poverty. The difficult policy issue in these circumstances is the following: What can be done to improve the economic conditions of women? The solution to rural poverty and food insecurity among women is to introduce technologies that will improve the economic returns of cassava production and processing. These measures include:

- the development and diffusion of labor-saving technologies for cassava field tasks, especially for harvesting;
- the development and diffusion of labor-saving methods for cassava processing, particularly the peeling task; and
- the diffusion of available labor-saving processing technologies to areas where such technologies are not available.

The COSCA study provides consistent evidence that when cassava is primarily a subsistence crop produced for home consumption, women are the dominant participants in production, harvesting, and processing. As the cassava transformation proceeds and cassava becomes a cash crop that is produced for urban centers, however, the workload of men increased for all tasks, from production to harvesting to processing. This important finding emerges from a comparison of the proportion of cassava work done on each task by men in four countries at different stages of the cassava transformation: the Congo, which is at the rural/urban food staple stage; Tanzania, which is at the famine-reserve stage; the Côte d'Ivoire, which is at the rural food staple stage; and Nigeria, which is at the cash crop for urban consumption stage (fig. 10.2). For each task, the proportion of work done by men was found to increase as the four countries progressed through the different stages of the cassava transformation. For example, the proportion of the cassava weeding done by men in Nigeria was higher than in the Côte d'Ivoire, while the proportion of the

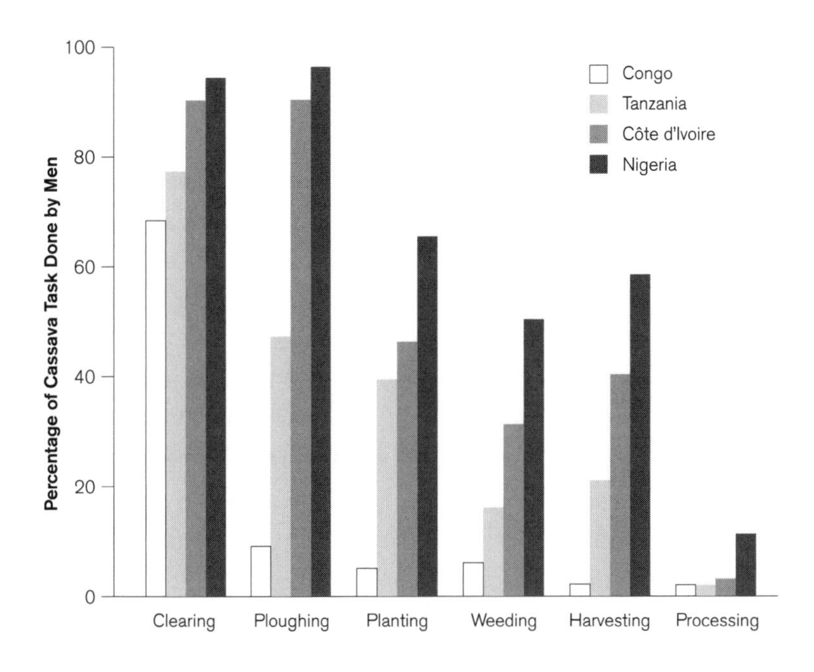


**Figure 10.2.** Percentage of Cassava Tasks Performed by Men in the Congo, Côte d'Ivoire, Tanzania, and Nigeria. *Source: COSCA Study.*

cassava weeding done by men was higher in the Côte d'Ivoire than in Tanzania, where cassava is used as a famine-reserve crop.

In countries such as Nigeria and Ghana, where the cassava transformation has advanced to the stage of a cash crop for urban consumption, men are staying at home to produce cassava side by side with women. As the cassava transformation progresses, men contribute more labor to each cassava production task. In most places, women assume increasing responsibilities in cassava production and processing as the tasks advance from field-based to home-based activities, such as processing, because women are able to combine these activities with homemaking activities. To summarize, men provided more than half of the labor on cassava fields in five of the six COSCA countries: the Côte d'Ivoire, Ghana, Nigeria, Tanzania, and Uganda.

## Access to Improved Technologies by Men and Women

Chapter 6 reported that the main inputs in cassava production in Africa are land, labor, planting materials, and, in some places, mechanized land clearing and plowing. The lack of access to farmland by women was not found to be a constraint on how much cassava women planted. Both women's and men's cassava fields were found on inherited, purchased, or rented farmland. Women are able to inherit, purchase, or rent farmland in most of the COSCA villages in the Congo, Ghana, and Tanzania, and in a large percentage of the study villages in the Côte d'Ivoire, Nigeria, and Uganda. In most of the villages in the Côte d'Ivoire, Ghana, Nigeria, Tanzania, and Uganda, men worked more, in proportionate terms, on women's cassava fields than women worked on men's cassava fields, either as family labor or as hired labor outside the household. Hence, women did not lack access to men's labor for their own cassava production.

Cassava planting materials—that is, the stem cuttings—were readily available to both men and women because they collected the cuttings from their own fields. The COSCA study found that although the agency responsible for the multiplication and distribution of the TMS varieties in Nigeria was 95 percent male, the TMS varieties were planted almost as frequently in women's fields (55 percent of women's cassava fields) as in men's fields (around 60 percent of men's cassava fields). During this same time period, the entire extension staff of the Ministries of Agriculture in the eastern states of Nigeria (82 percent male and 18 percent female) was engaged in the diffusion of TMS varieties.

The COSCA study found that the average cassava yield in the six countries was 11.0 tons per hectare on women's fields and 13.2 tons per hectare on men's fields. The lower yield in the women's fields was due to the harvesting of cassava in the women's fields in their compound garden at an earlier age of maturity than in men's fields. The average age of cassava fields at harvest in the six study countries was 15.7 months in men's fields and 12.9 months in women's fields. In addition, previous studies have shown that women are often allocated farmland of inferior soil fertility for their fields, further explaining the difference in the average yields between men's and women's cassava fields (Palmer-Jones 1991).

To summarize, in Nigeria, the TMS varieties were adopted as frequently in women's cassava fields as in men's cassava fields. But in Uganda, mechanized bush clearing and tillage were practiced more frequently in men's cassava fields than in women's fields because women's cassava fields are small and often located in compound gardens near their homes. The lack of access to farmland, labor, or extension advice was not found to be a constraint on cassava production by women.

### Cassava Processing and Marketing

In the six COSCA study countries, cassava processing was carried out mostly by women in 76 percent of the COSCA villages; equally by both women and the men in 20 percent of the villages; and mostly by men in only 4 percent of the villages (table 10.2).[4] Men carried out mostly the grating and pressing tasks in Nigeria and Ghana, where cassava grating and pressing tasks had been mechanized and cassava was a cash crop for urban consumption. In Nigeria, men were involved in cassava processing in more than 50 percent of the COSCA villages with mechanized graters that prepared *gari* for sale to urban consumers. In *gari* preparation, women manually peel and wash the cassava, men grate and press the cassava, then women toast the mash to make granulated products. This finding implies that men increase their labor inputs in cassava processing in areas where the crop is mainly produced for urban consumers.

In villages in five COSCA countries excluding the Côte d'Ivoire, men owned twice as many food-processing machines as women. In the Côte d'Ivoire, the number of women owning processing machines was almost ten times larger than the number of male owners. However, in all of these countries the services of the machines were available to both men and women on request. Therefore, although women generally owned fewer cassava-processing machines than men, they had easy access to these machines. We have shown that a large percentage of cassava fields in the COSCA study villages were planted for sale, especially in countries such

**Table 10.2.** Percentage of Villages Where Women and Men Carried Out Cassava Processing Tasks, Average for Six COSCA-Study Countries. *Source: COSCA Study.*

| PROCESSING TASK | WOMEN | MEN | MEN & WOMEN | TOTAL |
|---|---|---|---|---|
| | | PERCENTAGE | | |
| Peeling | 69 | 2 | 29 | 100 |
| Splitting | 79 | 3 | 18 | 100 |
| Crushing | 71 | 11 | 18 | 100 |
| Pressing | 79 | 5 | 16 | 100 |
| Sieving | 77 | 2 | 21 | 100 |
| Cooking | 86 | 2 | 12 | 100 |
| Other tasks | 88 | 3 | 9 | 100 |
| AVERAGE | 76 | 4 | 20 | 100 |

as Nigeria and Ghana where cassava is produced mainly as a cash crop for urban consumption. In all the COSCA countries, the proportion of men's cassava fields planted for sale was higher than the proportion of those of women.

In the COSCA countries, an average of 40 percent of the study villages sold their cassava mostly in the fields; 15 percent at home; and 45 percent in the marketplace. The sales were carried out mostly by women in 60 percent of the villages, mostly by men in 20 percent of the villages, and by women and men equally in 20 percent of the villages.

## Gender Surprises: Policy Implications

The COSCA studies have generated five findings that challenge the conventional wisdom that "cassava is a woman's crop" and that more female extension agents should be hired to assist women in increasing the productivity of the cassava food system. The first is that *both* men and women make significant contributions of their labor to the cassava industry in most of the COSCA countries. Women and men were found to specialize in different tasks, however. Men worked predominantly on land clearing, plowing, and planting, while women specialized in weeding, harvesting, transporting, and processing.

The second gender surprise is that women were found to contribute less than half of the total labor inputs in the cassava system in five of the six COSCA study countries. The exception was the Congo. Therefore, it is a vast overstatement to categorize cassava as " a woman's crop." In fact, both men and women play strategic but changing roles during the cassava transformation process.

The third gender surprise is that as the cassava transformation proceeds and cassava becomes a cash crop produced primarily for urban centers, men increase their labor contribution to each of the production and processing tasks. For example, the proportion of the cassava weeding done by men was higher in Nigeria than in Tanzania because Nigeria is at the cash crop stage of the cassava transformation while Tanzania is still at the famine-reserve stage.

The fourth gender surprise is the COSCA finding that the introduction of labor-saving technologies in cassava production and processing has led to a redefinition of gender roles in the cassava food system. In Nigeria, on farms where land clearing or plowing was mechanized, men increased their labor inputs in planting, weeding, and harvesting. As the processing tasks become mechanized in Nigeria, the contribution of male labor to cassava processing increases, because men operated all of the processing machines. When processing became mechanized, women shifted their labor to production tasks, such as weeding, while men managed the mechanized processing tasks. Both the third and the fourth findings challenge the validity of the claim that women's workload increases relative to men's as commercialization of agriculture proceeds.

The fifth gender surprise is that women who wanted to plant cassava were not constrained by the lack of access to new cassava production technologies or to essential production inputs such as farmland, cuttings of the TMS varieties, and hired labor. Both women and men were able to plant cassava on land that they had inherited, purchased, or rented. Finally, although most of the cassava-processing machines were owned by men in most of the COSCA countries, both men and women had access to the machines by paying a fee to the machine owners.

What policy implications flow from these gender surprises? Although there are persistent calls by donor agencies and gender experts on African policy makers to hire more women agricultural extension workers in cassava-growing areas, COSCA research shows that *both* men and women play significant and changing roles in the cassava food system in Africa.

Country-specific studies are needed to determine appropriate policy measures based on the stage of the cassava transformation and the market outlook for cassava for consumers and livestock feed and industrial uses. For example, in Nigeria and Ghana, where cassava is produced as a cash crop, more men are staying in the villages and working in cassava production and *gari* processing. Similarly, in the Congo, men are staying in tree crop-producing villages and working on cassava for home consumption because they earn higher cash income from tree crops than from employment in urban areas. Yet outside the tree crop areas, young men migrate out of the Congolese villages to escape rural poverty. Women are therefore left to produce food for their families under conditions of extreme rural poverty.

The solution to poverty and low returns to both men and women engaged in cassava production and processing is to take a holistic approach and introduce improved varieties and processing technologies to improve the economic returns of cassava production and processing and drive down the real (inflation adjusted) price of cassava in rural and urban markets. The specific measures to be taken include:

- development and diffusion of labor-saving technologies for cassava field tasks, especially for harvesting;
- development and diffusion of labor-saving methods for cassava processing, particularly the peeling task;
- diffusion of labor-saving processing technologies to regions and countries where such technologies do not exist; and
- increasing "market pull" by improving roads and increasing the access of cassava farmers and *gari* producers to both rural and urban markets.

CHAPTER 11

# Consumption

## Introduction

Cassava is a major source of calories for roughly two out of every five Africans.[1] However, many international agencies and bilateral donors are hesitant to extend loans and grants to African nations to help them increase the production of root crops such as cassava because of the long held wrong belief that cassava is an "inferior good," that is, that the per capita consumption of cassava declines as per capita income increases. For example, soon after International Food Policy Research Institute (IFPRI) was established in 1975, it reported that "since these root crops require much larger bulk to provide calories than do cereals, and are low in protein, demand may shift towards cereals as has occurred in other countries" (International Food Policy Research Institute 1976, 35). Today, the low status accorded cassava by the international organizations and donor agencies flows from two misleading myths: that cassava is an inferior food that is produced by and for rural households and that because of its low protein content cassava is a nutritionally inferior food crop. However, the IFPRI recently concluded that root crops such as cassava are important for smallholders in the marginal areas of Africa, Asia, and

South America and that special steps should be taken to boost cassava production, especially in Africa (Pinstrup-Anderson, Pandya-Lorch, and Rosegrant 1999). Yet much of the future growth of the cassava industry will depend on whether investments are going to be forthcoming to make cassava competitive with rice and wheat in urban centers.

This chapter examines cassava consumption patterns over time and marshals evidence to show: (1) that cassava is a major source of calories for a large proportion of the population in Africa; (2) that from 1961 to 1998 total cassava consumption more than doubled in Africa, compared with only about a 20 percent increase in South America (FAOSTAT); (3) that cassava waste is high, especially because of inefficient processing methods, and (4) that the income elasticities of demand for cassava and cassava products such as *gari* are high among rural and urban households in Ghana and Nigeria and among rural households in Tanzania and Uganda. The chapter concludes that cassava and cassava products, such as *gari,* are an important food staple with favorable market prospects for low-, medium-, and high-income households in Africa.

## Cassava Consumption Patterns

Cassava production in Africa is used almost exclusively for consumption as food. In fact, 95 percent of the total cassava production, after accounting for waste, was used as food in Africa in the late 1990s.[2] By contrast, during the same period, 55 percent of total cassava production in Asia and 40 percent of that in South America were used as food.

In Africa, total cassava consumption has more than doubled in the last thirty years, from twenty-four million tons per year from 1961 to 1965 to fifty-eight million tons per year from 1994 to 1998, after accounting for waste (FAOSTAT). The large increase in the total cassava consumption in Africa is due to a significant increase in per capita consumption in countries such as Nigeria and Ghana where cassava is produced as a cash crop for urban consumption. For example, the FAO data show that in Ghana, from 1961 to 1998, per capita cassava consumption nearly doubled, from 130 kilograms per person per year from 1961 to 1965 to 256 kilograms

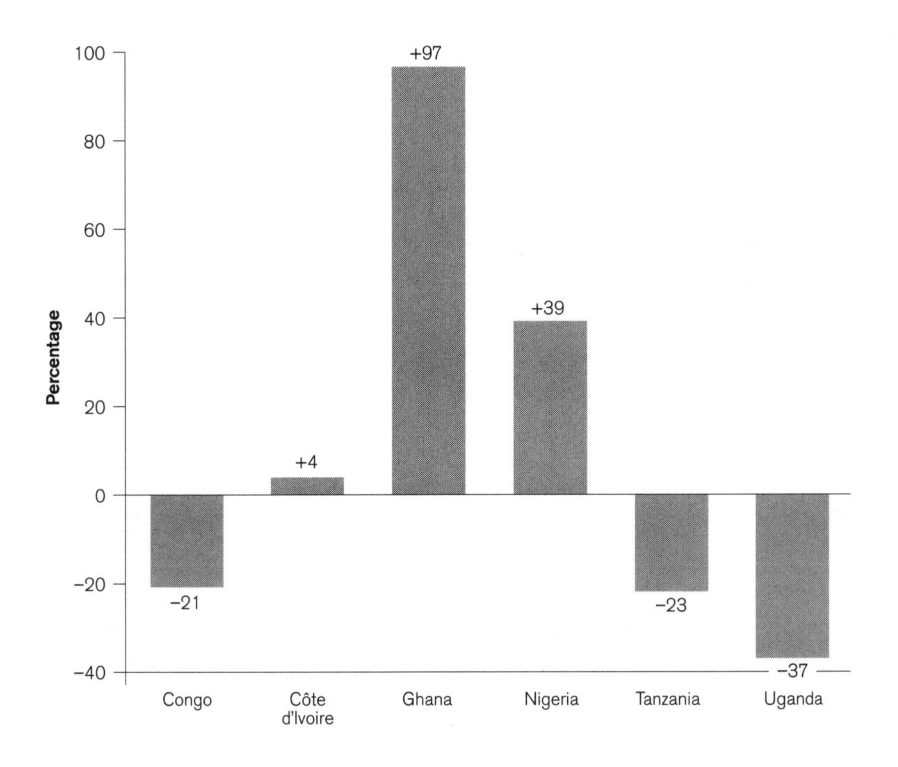


**Figure 11.1. Percentage Change in Per Capita Consumption of Cassava in the COSCA Countries from 1961 to 1998.** *Source: FAOSTAT.*

per person per year from 1994 to 1998. In Nigeria, the per capita consumption increased by 40 percent, from 88 kilograms per person per year from 1961 to 1965 to 120 kilograms per person per year from 1994 to 1998 (FAOSTAT). Yet the per capita cassava consumption in the same period stagnated in the Côte d'Ivoire and declined significantly in the Congo, Tanzania, and Uganda (fig. 11.1). The availability of cassava in a convenient food form, such as *gari,* played a major role in the increase in per capita cassava consumption in Nigeria and Ghana. Future increases in cassava consumption in other African countries will depend on how well cassava is prepared into food forms which make it an alternative to wheat, rice, maize, and sorghum for urban consumers.

The FAO data also show that on an average basis, cassava roots are the single largest source of calories in seven countries with a total population of 240 million, or 40 percent of the population of Africa, from 1994 to 1998.[3] In these seven countries, cassava contributed an average of 590 calories per person per day. In another eleven countries with 23 percent of Africa's population, cassava was the second-largest source of calories.[4] In these countries, cassava provided an average of 311 calories per person per day from 1994 to 1998.

However, these averages underestimate the importance of cassava in specific countries. In the Congo, for example, many families eat cassava for breakfast, lunch, and dinner. From 1994 to 1998, cassava contributed over 1,000 calories per person per day or about 55 percent of the average daily calorie intake in the Congo (FAOSTAT). Furthermore, the FAO data do not account for the consumption of cassava leaves. Yet the COSCA study reveals that cassava leaves are widely consumed as a vegetable in several places where cassava is grown, including the Congo and Tanzania. Since cassava leaves are rich in protein, vitamins A and C, and some minerals (iron and calcium), they partially compensate for the shortage of these nutrients in the roots (Latham 1979, 172).

Fresh roots of sweet cassava varieties are eaten raw, boiled, or fried in oil. Often the boiled fresh cassava roots are pounded into a paste with a mortar and pestle and eaten as pasty dough balls with protein-rich sauce. The consumption of boiled and pounded fresh cassava roots with a sauce of oil-palm fruit and goat meat is a popular lunch among urban workers in Ghanaian cities such as Accra.

Processed cassava food products are eaten as pasty porridge or dough balls with a seasoned sauce. Bits of the porridge or dough balls are dipped into the sauce and eaten. Ingredients of the sauce vary greatly, depending upon the availability of vegetables, meat, fish, melon seeds, peas, peppers, and other spices. Palm oil is a standard ingredient of the sauce in the forest zone, while in the transition zone palm oil is replaced by shea butter or peanut cake (Johnston 1958; Jones 1959; Grace 1977). In places where cassava is consumed every day, variation in the diet is achieved by varying the ingredients of the sauce.

Texture was found to be one of the three common quality characteristics (color, texture, and sour taste) of cassava products desired by consumers in the COSCA study villages. Texture is an important consideration in cassava products because there is an art to molding pasty foods into balls with one's fingers, dipping the balls in a sauce, and swallowing, sometimes without chewing. The texture is important because it must ensure that food balls do not break up. The foods eaten in the form of balls are called *eba* or *fufu* (*foufou*) in West and Central Africa, *ugali* in East Africa, and *nsima* in Southern Africa. *Eba* is made from *gari; fufu* is made from cassava, yam, and plantain; and *nsima* and *ugali* are made from grain. Cassava dried root flour is often mixed with grain to make *nsima* or *ugali.*

## Pre- and Postharvest Losses

High pre- and postharvest losses are a serious problem in Africa. In the late 1990s, FAO data show that an estimated 28 percent per year of total cassava production in Africa was classified as waste (FAOSTAT).[5] Waste is caused by inefficient processing methods and poor market infrastructure.

The traditional processing methods are inefficient in carrying out most of the processing operations. Hand-peeling in particular, is labor-intensive. Smaller roots are often discarded because they are more difficult to peel except in the case of an acute food shortage. Roots that are damaged or broken during harvesting or transporting and roots with irregular shapes that are too difficult to peel are also often discarded. The proportion of roots considered "usable" depends on the variety, yield, and food security status (Fresco 1986).

Since cassava roots begin to deteriorate within forty-eight hours after harvest, farmers usually take roots to the market or to the processing center as soon as they are harvested. Yet the crop cannot be harvested faster than the available markets can absorb the produce. Rwandan farmers have been observed to abandon cassava roots in the marketplace because transportation costs back to their homes exceeded the value of the unsold cassava roots (Ndamage 1991).

A loss in quality is associated with a delayed harvest. The production of roots suitable for consumption is usually greatest between twelve and eighteen months after planting. Yet many varieties will continue to grow for up to forty-eight months, and starch production continues to increase. As the roots grow older, however, they become tougher and woodier and are therefore harder to prepare for human consumption (Jones 1959).

In countries such as Nigeria and Ghana where a large percentage of cassava is produced as a cash crop for urban consumption, cassava is usually harvested early, that is, between nine and twelve months after planting. When prices are high, farmers will harvest many roots that would ordinarily be considered immature. By contrast, in countries where cassava is produced as a famine-reserve crop, such as in Tanzania, or where it is produced as a rural food staple, such as in the Congo, the Côte d'Ivoire, and Uganda, cassava is harvested late most of the time. As an example, to reduce losses that result from cassava root rot, farmers in the Congo were advised by the national extension agency to shorten the period that cassava was in the ground to between twelve and twenty-four months. However, this advice was not widely accepted. Farmers who attempted to harvest at between twelve and twenty-four months encountered problems in transporting the harvested cassava to the urban centers because of poor road conditions. Most of the roads in the Congo are impassable at certain times of the year (Makambila 1981).

Waste can be reduced by improving cassava-processing methods, which will result in products that have less bulk and a longer shelf life. A cassava product that is not bulky and has a long shelf life can be transported at a reduced cost over poor roads to distant urban markets and stored until sold.

## Food Expenditure Patterns

The COSCA study shows that in the main cassava-producing areas, 85 percent of the rural households in the Côte d'Ivoire, 87 percent in Ghana, 80 percent in Nigeria, 62 percent in Tanzania, and 88 percent in Uganda prepared and ate a cassava meal at least once in the week before the

**Table 11.1.** Percentage of Rural Households that Prepared and Ate Meals of Major Staples in a One-Week Interval during the COSCA Study in the Côte d'Ivoire, Ghana, Nigeria, Tanzania, and Uganda. *Source: COSCA Study.*

| STAPLE | CÔTE D'IVOIRE | GHANA | NIGERIA | TANZANIA | UGANDA |
|---|---|---|---|---|---|
| Cassava | 85 | 87 | 80 | 62 | .88 |
| Grains | 38 | 81 | 78 | 66 | 69 |
| Rice | 75 | 43 | 77 | 51 | 18 |
| Pulses | 3 | 42 | 79 | 66 | 89 |
| Banana or plantain | 63 | 33 | 28 | 42 | 55 |
| Yam | 40 | 66 | 68 | 3 | 4 |
| Sweet potato | 0 | 5 | 12 | 28 | 48 |

households were interviewed. The cassava-based meal was the most frequently eaten meal in rural households in the cassava-producing areas in the COSCA countries (table 11.1). In some countries, cassava was consumed daily, sometimes more than once a day. Meals based on maize, millet, sorghum, rice, banana, yam, and pulses were also consumed at varying frequencies, depending on country, village, and household.

The COSCA researchers found that the share of convenient forms of cassava, such as *gari* and *attieke,* increase in the household food budget as the household's income increases.[6] For example, in Nigeria, the COSCA study found that the share of *gari* in the household's food budget was 0.6 percent among the low-income households, as compared with 3.2 percent among the high-income households. The COSCA study further shows that the trend is similar in the Côte d'Ivoire, where a cassava granule, *attieke,* is an important food staple (table 11.2). In Ghana, a household economic survey found that the share of *gari* in the household's food, beverages, and tobacco budget was higher in the urban centers (1.11 percent) than in the rural areas (0.79 percent) (Central Bureau of Stastistics 1979). In Nigeria and the Côte d'Ivoire, the high-income households prefer *gari* more than do households in the low-income group. In Ghana, urban dwellers prefer *gari* more than do rural households. *Gari* consumption is low among low-income households because it is more expensive than fresh and dried roots.

By contrast, the share of dried cassava roots in the household food budget decreases as household income increases. For example, in Nigeria, the COSCA study found that the share of dried cassava roots in the

**Table 11.2.** Percentage of Rural Households' Food Expenditure on Dried Cassava Roots and Granules in Lowest and Highest Income Quartile in the Côte d'Ivoire, Ghana, Nigeria, Tanzania, and Uganda. *Source: COSCA Study.*

| COUNTRY | DRIED ROOTS | | GRANULES[a] | |
|---|---|---|---|---|
| | INCOME GROUP | | INCOME GROUP | |
| | LOWEST QUARTILE | HIGHEST QUARTILE | LOWEST QUARTILE | HIGHEST QUARTILE |
| | PERCENTAGE | | PERCENTAGE | |
| Côte d'Ivoire | 6.2 | 0.2 | 5.4 | 6.0 |
| Ghana | 4.2 | 3.8 | – | – |
| Nigeria | 12.8 | 5.4 | 0.6 | 3.2 |
| Tanzania | 13.7 | 7.1 | – | – |
| Uganda | 6.7 | 4.5 | – | – |

[a]*Gari* in Nigeria and *attieke* in the Côte d'Ivoire.

household's food budget was 12.8 percent among low-income households, as compared with 5.4 percent among the high-income households. The study also found that the trend was similar in the Côte d'Ivoire, Ghana, Tanzania, and Uganda. In Ghana, the household economic survey showed that the share of dried cassava roots in the households' food, beverages, and tobacco budgets was higher in the rural areas (0.75 percent) than in urban centers (0.61 percent) (Central Bureau of Stastics 1979). In the Côte d'Ivoire, Nigeria, Tanzania, and Uganda, the low-income households are more dependent on the cheaper dried roots than are households in the high-income group. In Ghana, rural households are more dependent on the dried roots than are the urban dwellers.

Cassava is appealing to low-income households because it is the cheapest source of food calories. Compared with grains, fresh and dried cassava roots are very cheap sources of calories. Calories are significantly cheaper from fresh roots of sweet cassava varieties than from maize in various rural village market centers in Nigeria (table 11.3). Similarly, calories derived from dried cassava roots are significantly cheaper than those derived from maize in various rural market centers in Ghana.

The consumption of cassava in form of *gari* increases as household income increases. This underscores the need to promote and diffuse the

**Table 11.3.** Nigeria and Ghana: Retail Price of 1,000 Calories from Fresh Roots of Sweet Cassava Varieties, Dried Roots, and Maize in Rural Market Centers, 1992. *Source: COSCA Study.*

| **NIGERIA** (NAIRA/1000 CALORIES) | | | **GHANA** (CEDIS/1000 CALORIES) | | |
|---|---|---|---|---|---|
| RURAL MARKET CENTER | FRESH CASSAVA ROOTS | MAIZE | RURAL MARKET CENTER | DRIED CASSAVA ROOTS | MAIZE |
| Donga | 0.36 | 0.95 | Sagboi | 34 | 49 |
| Garbabi | 0.38 | 0.85 | Tafiano | 35 | 53 |
| Suwabarki | 1.09 | 1.37 | Nkurakan | 44 | 71 |
| Guyuki | 0.85 | 1.60 | Koluedor | 32 | 83 |
| Namtarigure | 0.80 | 1.20 | | | |
| Yaburawa | 0.63 | 1.11 | | | |
| Wuse | 0.81 | 1.07 | | | |
| Busanfung | 0.71 | 3.20 | | | |
| Ofabe | 0.24 | 0.60 | | | |

Note: 1.00 Naira = $US0.06, 430 Cedi = $US1.00.

use of *gari* technology in Tanzania, Uganda, and other countries in order to make cassava attractive to medium- and high-income rural and urban consumers. The future of cassava as a rural and urban food staple will depend on the cassava's ability to compete with grain in terms of cost, convenience, and availability in urban markets. This can be achieved by the diffusion of labor-saving technologies for cassava production, harvesting, and processing, particularly peeling.

### Income Elasticity of Demand

Many commentators on food policy assume that cassava is an inferior good, that is, that cassava consumption declines with rising incomes. Yet until the COSCA study was carried out in the late 1980s and early 1990s, there was little empirical data to measure the income elasticity of demand for cassava.

The income elasticity of demand provides insight into the level of market demand for a commodity. This function measures the percentage of change in the quantity of a commodity purchased (consumed) by consumers in response to a 1 percent change in their incomes. A negative income elasticity of demand means that the quantity of the commodity purchased by consumers will decline with rising incomes. A zero income elasticity of demand means that the amount of the commodity demanded will be unchanged with rising incomes. An income elasticity of demand between zero and one implies that a 1 percent increase in incomes will cause consumers to increase the amount of the commodity they are willing to purchase, although by less than 1 percent. Finally, an income elasticity of demand of more than one implies that market demand is very high for the commodity. This is because a 1 percent increase in income will cause consumers to increase the amount of the commodity they are willing to purchase by more than 1 percent. Scholars and policy makers who dismiss cassava as an inferior good assume that the income elasticity of demand for cassava is negative or zero.

However, the COSCA consumption data reveal that the income elasticity of demand estimates for cassava products among rural households in Nigeria, Tanzania, and Uganda were all greater than zero, and in some cases they were greater than one (table 11.4).[7] Surprisingly, the estimates for cassava were about the same as the estimates for maize among the same households in the same countries. Among individual cassava food products, fresh cassava roots had a higher income elasticity of demand than *gari* and dried cassava roots, except in Uganda. In Uganda, dried roots had a higher income elasticity of demand estimate than fresh cassava roots, because dried cassava food products are less bulky and have a longer shelf life, and therefore are less expensive to transport, than fresh cassava roots. In countries such as Nigeria and Tanzania, where bitter cassava varieties are more widely produced than the sweet varieties, the sweet types are more commonly grown and eaten as raw or boiled vegetables around market centers, where household incomes are generally higher than in remote villages. In Uganda, sweet cassava is more common, irrespective of distance to market centers.

In Nigeria, the income elasticity of demand estimate for *gari* was significantly higher than that for dried roots. The income elasticity estimate for *gari* was also higher than that for maize, even among

**Table 11.4.** Income Elasticity of Demand for Cassava and Other Food Staples for Rural Households in Nigeria, Tanzania, and Uganda; Rural and Urban Households in Ghana. *Source: 1. COSCA Study; 2. Alderman 1990.*

| STAPLE | NIGERIA[1] | | | TANZANIA[1] | | | UGANDA[1] | | | GHANA[2] | |
|---|---|---|---|---|---|---|---|---|---|---|---|
| | ALL SAMPLE HOUSE-HOLDS | LOW INCOME HOUSE-HOLDS | HIGH INCOME HOUSE-HOLDS | ALL SAMPLE HOUSE-HOLDS | LOW INCOME HOUSE-HOLDS | HIGH INCOME HOUSE-HOLDS | ALL SAMPLE HOUSE-HOLDS | LOW INCOME HOUSE-HOLDS | HIGH INCOME HOUSE-HOLDS | RURAL HOUSE-HOLDS | URBAN HOUSE-HOLDS |
| All cassava | 0.78 | 0.84 | 0.76 | 0.77 | 0.80 | 0.66 | 1.00 | 1.00 | 1.00 | 0.73 | 1.46 |
| Fresh roots | 1.24 | 1.28 | 1.21 | 0.79 | 0.79 | 0.67 | 0.95 | 0.96 | 0.94 | – | – |
| *Gari* | 0.85 | 0.85 | 0.77 | – | – | – | – | – | – | – | – |
| Dried roots | 0.55 | 0.57 | 0.53 | 0.75 | 0.80 | 0.66 | 1.17 | 1.15 | 1.13 | – | – |
| Maize | 0.71 | 0.74 | 0.65 | 0.98 | 0.98 | 0.97 | 0.91 | 0.85 | 0.91 | 0.84 | 0.83 |
| Rice | 1.12 | 1.13 | 1.13 | 1.14 | 1.25 | 1.11 | 1.36 | 1.59 | 1.25 | 1.00 | 1.50 |
| Pulses | 1.02 | 1.01 | 1.02 | 1.06 | 1.05 | 1.07 | 0.78 | 0.84 | 0.69 | – | – |
| Banana and plantains | 2.06 | 1.97 | 1.69 | 0.94 | 0.93 | 0.95 | 1.22 | 1.38 | 1.16 | 0.13 | 1.10 |
| Yam | 0.91 | 0.90 | 0.92 | – | – | – | – | – | – | – | – |
| Sweet potato | – | – | – | 0.97 | 0.97 | 0.97 | 0.74 | 0.79 | 0.64 | – | – |

high-income rural households. The COSCA estimates show that cassava has as much market demand potential as maize.

In southeastern Nigeria, among the high-income rural and urban households, the income elasticity of demand estimate for *gari* (0.78) was higher than the estimate for rice (0.62) (Nweke et al. 1994). In Ghana, the income elasticity of demand estimates based on Living Standards Surveys data are equally surprising: the estimate for cassava was significantly greater than one among urban households (1.46) but less than one among rural households (0.73). Among the urban households, the estimate for cassava was about the same as the estimate for rice (1.50) but significantly greater than the estimate for maize (0.83) (Alderman 1990).

The income elasticity of demand estimates show that cassava is *not* an inferior food. These new findings provide convincing evidence that demand for cassava will continue to rise as income increases in the cassava-producing countries studied by COSCA. In fact, some cassava food products, such as *gari,* have a higher income elasticity of demand than some grains. Yet these surprising findings are not a call for complacency. In order for cassava to remain competitive with grain in urban diets, critical investments must be made to drive down processing costs and to carry out research by food scientists on new cassava products. The future market for cassava will depend on whether the quality and variety of cassava food products can be improved to make them attractive to low-, medium-, and high-income consumers in rural and urban centers.

### Summary and Conclusions

In Africa, over 90 percent of cassava production is consumed as food. Between 1961 and 1998, total cassava consumption tripled in Africa because of a large increase in per capita consumption in Nigeria and Ghana, where cassava is produced as a cash crop for urban consumption. The availability of cassava in a convenient food form, such as *gari,* played a major role in the increase in per capita cassava consumption in Nigeria and Ghana. The increase in per capita cassava consumption in other African countries will depend on how well cassava is prepared into food forms that make it as acceptable as grains in terms of quality and cost to urban consumers.

In some countries, such as the Congo, cassava accounts for more than 50 percent of daily calorie consumption. In several countries, including the Congo and Tanzania, cassava leaves are consumed as a vegetable. Cassava leaves are rich in protein, vitamins, and minerals. Everywhere in Africa where cassava is consumed, it is invariably consumed with a sauce that is made with ingredients rich in protein, vitamins, and minerals.

Cassava was found to be the cheapest source of calories among all food crops in each of the six study countries. As family incomes increased, the consumption of cassava as dried root flour declined while consumption in convenient food forms such as *gari* increased. Dried cassava root flour is cheaper than *gari* because of the high cost of processing *gari*. Medium- and high-income families were found to consume *gari* because it is cheaper and more convenient to cook than grains. The future of cassava as a rural and urban food staple will depend on the cassava's ability to compete with wheat, rice, maize, sorghum, and other grains in terms of cost, convenience, and availability in urban markets. Cassava can retain its competitive edge only through investments in labor-saving production, harvesting, and processing technologies.

The income elasticity of demand estimates for cassava products was found to be positive at all income levels. For some cassava products, the income elasticity of demand was above one. For example, in Nigeria the income elasticity of demand for *gari* was significantly higher than that for maize at all income levels. Among high-income urban households, the income elasticity of demand for *gari* was also higher than that for rice. In Ghana, the income elasticity of demand for cassava (all products combined) was higher among urban than among rural households. These estimates provide convincing evidence that cassava has a strong market demand in the COSCA study countries.

The conclusion that emerges from the data presented in this chapter is that cassava is not an inferior food. Rather, in certain forms, cassava is superior to some grains. *Gari* has substantial market demand among low-, medium-, and high-income rural and urban households. Yet the degree to which the future market demand for cassava for food consumption can be expanded will depend largely on the extent to which the quality and variety of cassava food products can be improved to make them attractive to a range of consumers in rural and urban centers.

The steps needed to reduce the cost of making cassava available to consumers in convenient food forms include:

- development and diffusion of labor-saving technologies for cassava field tasks, especially harvesting;
- development and diffusion of labor-saving methods for cassava processing tasks for which such methods have not been developed, particularly the peeling task; and
- diffusion of existing cassava processing technologies in areas where such technologies are not available.

# New Uses for Cassava: A Nigerian Case Study

## Introduction

We pointed out in chapter 1 that Nigeria is the most advanced of the six COSCA countries in the cassava transformation (table 1.1). Nigeria is now poised to move to the next stage, namely the feed and industrial raw material stage. Yet the transformation will not continue unless new markets are identified to absorb the increase in cassava production. The wide-scale adoption of TMS varieties and the resulting increase in yields have shifted the problem of the Nigerian cassava industry from supply (production) to demand issues, such as finding new uses for cassava in livestock feed and industries. Yet efforts to improve production and yields of tropical starch crops such as cassava often result in excess supplies for existing market demand, which, in turn, discourages production (Satin n.d.). For example, O. A. Edache, the director of the Federal Department of Agriculture in Nigeria, lamented in early 2001 that cassava producers were losing money because of a glut in the market and declining cassava prices.[1]

Nigeria has been chosen as a case study for new uses for cassava because it is the largest cassava producer and because its cassava transformation is the most advanced in Africa. We shall examine what needs to be done to increase the use of cassava in livestock feed and food manufacture; as starch in the textile, petroleum drilling, pharmaceutical, and soft drink industries; and as dried roots in the beer and alcohol/ethanol industries. This chapter is based on preliminary information collected in an exploratory survey in Nigeria in early 2001.

## Expanding the Use of Cassava in the Livestock Industry

In Nigeria, the proportion of cassava used in the livestock industry increased after the government banned the importation of maize in 1985–86 and feed mills were forced to use local raw materials such as cassava. After the import restrictions on maize and other crops were removed in 1995, there was no incentive for feed millers to reduce the quantity of cassava they were using, because cassava is cheaper than maize, the feed mills had modified their plants to mill dried cassava roots, and farmers were getting higher yields from TMS varieties.

In Nigeria, the 5 percent of total cassava production that is used as feed is significantly lower than in Brazil (50 percent) because in Nigeria, cattle, sheep, and goats are free grazed and pigs rummage on household waste. Also, the poultry industry has only 125 million birds in Nigeria, as compared with 867 million in Brazil in 1998 (FAOSTAT).[2] Because of the pessimistic outlook for a major increase in the amount of cassava fed to livestock in Nigeria, the logical next step is to examine the global outlook for Nigerian cassava exports for livestock feed. The key country to examine is Thailand, because it has dominated the export of cassava pellets for livestock feed for more than three decades. In Thailand, only a small percentage of national cassava production is consumed as food. The most important uses for cassava are for livestock feed and starch.

The production process for cassava pellets in Thailand is shown in figure 12.1. After sand and impurities are removed, the dried cassava roots are ground with a hammer mill. Cassava particles together with steam are forced through holes in a die. The compressed material emerges hot from the other side of the die. After cooling, the strands are cut to length to produce pellets (Ratanawaraha, Senanarong, and Suriyapan 1999, 16).

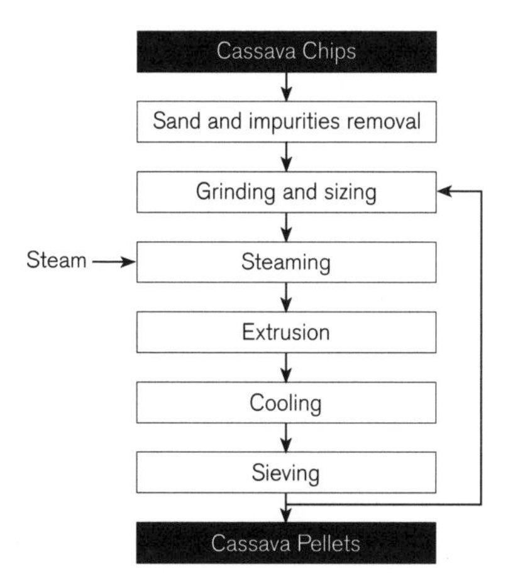


**Figure 12.1. Process for Production of Cassava Pellets.** *Source: Ratanawaraha, Senanarong, and Suriyapan 1999.*

Beginning in the 1960s, the government of Thailand encouraged private firms to set up pellet factories and produce cassava pellets for export to the European Union (EU). The private sector responded, and pellet exports literally "took off." In fact, exports increased from 100,000 tons in 1966 to a peak of 9 million tons in 1989. Yet because of competition with U.S. grain exports to the EU market, the price of cassava pellets has declined, making it unattractive for Thailand to produce cassava for export. Also the EU has set an annual quota for Thai cassava imports of 5.25 million tons. However, because of low prices, Thailand did not meet its export quota from 1994 to 1998. In fact, Thai pellet exports have declined from 9 million tons in 1989 to 3 million tons in 1998 (Ratanawaraha, Senanarong, and Suriyapan 1999, 18). There are currently two hundred pelletizing factories in Thailand, with a total capacity of 10 million tons per year. Yet because of depressed prices, they are operating at only 50 percent of capacity.

What is the outlook for Nigerian pellet exports? Faced with overcapacity in pellet factories in Thailand and depressed world prices of cassava pellets, the answer is clear: Nigeria should concentrate on expanding the use of cassava in livestock feed at home rather than trying to break into the EU market at this time. What can be done to increase the use of cassava in livestock feed in Nigeria? A poultry feed trial has shown that if cassava roots and leaves were combined in a ratio of four to one, the mixture could replace maize in poultry feed and reduce feed cost without a loss in weight gain or egg production (Tewe and Bokanga 2001). If this important research finding is diffused and adopted by farmers and livestock feed producers, the amount of cassava used in livestock feed in Nigeria would increase and cassava leaves could become an additional source of income for cassava producers. At present, cassava leaves have no market value in Nigeria because they are not consumed as a vegetable as they are in the Congo and Tanzania.[3]

### Expanding the Use of Cassava in Food Manufacturing

Technologies exist for the use of cassava as a partial substitute for wheat in bread-making, and biscuits, pastries, and snack foods manufacture (Satin 1988; Eggleston and Omoaka 1994; Defloor 1995; and Onabolu et al. 1998). But in Nigeria in the late 1990s, an insignificant 3 tons of cassava was used per year for food manufacture compared with maize, 133,000 tons (FAOSTAT). Use of cassava as a partial substitute for wheat in food manufacture will increase if the practice can result in a reduction in the prices of the manufactured composite cassava and wheat flour food products compared with the prices of the same products made with 100 percent wheat flour.

But in Nigeria, because of an array of reasons, the composite cassava and wheat flour food products are not cheaper than the 100 percent wheat flour food products. For example, a partial substitution of cassava for wheat in bread flour requires expensive supplementary viscosity enhancers such as eggs, milk, and gums to compensate for the lack of gluten in cassava (Eggleston and Omoaka 1994; Defloor 1995; and Onabolu et al. 1998).

Using cassava flour for bread-making and for the manufacture of biscuits, pastries, and snack foods requires a reliable supply of cassava flour with constant quality (Delfoor 1995). But in Nigeria, cassava flour available in the market varies widely in quality because of the wide range of traditional methods of preparation used. A cassava flour of standard high quality will be more expensive and will increase the cost of the food products.

Other important factors such as the cassava variety, age of the cassava root, and the cassava growing environment also influence the quality of the food products in which cassava flour substitutes partially for wheat flour (Eggleston and Omoaka 1994; Defloor 1995). Measures to standardize cassava varieties, age of cassava roots, and the cassava growing environments will further increase the costs of the food products in which the cassava flour is used to substitute partially for wheat flour.

In Nigeria, the technologies for the use of cassava as a partial substitute for wheat flour in food manufacture that were developed at the IITA and other research centers were adopted by food industries when wheat flour was made scarce by the wheat import ban in the mid-1980s. But when the wheat import ban was removed in the mid-1990s, the food industries readily reverted to the use of 100 percent wheat flour in food manufacture (Bokanga and Tewe 1998).

In Nigeria, increase in the use of cassava in food manufacturing industries does not depend on technologies for partial substitution of cassava flour for wheat flour in bread, cookies, pastries, etc. But increase in the use of cassava in food manufacturing industries requires the development of technologies for industrial manufacture and packaging of traditional African cassava food products which have a snack value such as *gari, attieke,* and *chickwangue*.

We explained earlier that in Africa, past attempts to manufacture *gari* and *chickwangue* industrially failed because they were unprofitable. But in Brazil, recent development in the use of cassava in food industries shows that sustained investment in research and development can make industrial manufacture of a 100-percent-cassava-starch traditional food product profitable. For example, the main cassava-based fast food in Brazil is *pao de queijo*, a type of bread made with sour cassava starch, which has been fermented and dried (Vilpoux and Ospina 1999).

In Brazil, the preparation of *pao de queijo* by traditional methods has similar problems as the preparation of *gari, attieke,* and *chickwangue* in Africa. For example, in Minas Gerais, one of the traditional *pao de queijo* production states in Brazil, almost every family has its own recipe. Sour starch gives *pao de queijo* a very acid taste, which is appreciated in Minas Gerais state. In other states however, where consumption is more recent, people prefer *pao de queijo* with a mild taste. Industrial manufacture of *pao de queijo* needs a standard product with consistent quality. These criteria are difficult to maintain in *pao de queijo* because of the low and unstable quality of sour starch. Making sour starch is labor-intensive. The sour starch industry is competitive with a large number of small producers. The small capacity limits the possibility to access modern technologies and market information (Vilpoux and Ospina 1999).

But through sustained investment in research and development in Brazil, *pao de queijo* was transformed from a small-scale homemade product to a large-scale factory-manufactured product by first replacing fermentation and sun-drying with a chemical process in making the sour starch. Later, the sour starch was replaced with other types of cassava starch that are acceptable to consumers. Since the early 1990s, the preparation of *pao de queijo* by small-scale manufacturers in the traditional production states such as Minas Gerais has been declining while larger more modern companies are expanding in other states. The entrance of large companies changed most of the industry. The operators of the large companies have better education, better access to new technologies, and market information (Vilpoux and Ospina 1999).

In Brazil, the research and development in the improvement of *pao de queijo* were carried out mostly by the private sector. But the expansion in the consumption of the *pao de queijo* was facilitated by political support. For example, the consumption was endorsed by the former Brazilian president, Itamar Franco. He required that *pao de queijo* be present at all official meetings. Since the mid-1990s, Brazilian consumption of *pao de queijo* has increased greatly, changing from a regional to a nation-wide fast food. It is also possible to find *pao de queijo* in other South American countries such as Argentina and Peru (Vilpoux and Ospina 1999).

To summarize, there are technologies for use of cassava as a partial substitute for wheat in food manufacture. But in Nigeria, cassava is not used for food manufacture because food products made with 100 percent wheat flour are cheaper and preferred by consumers. Brazilian examples show that sustained investment in research and development on industrial manufacture of African cassava food products such as *gari, attieke,* and *chickwangue,* which have snack value, can lead to increased use of cassava in food manufacturing industries.

## Expanding the Use of Cassava Starch As an Industrial Raw Material

Plant starches such as wheat, rice, corn, potato, and cassava are used as an industrial raw material in direct form or in a number of derivative forms.[4] We shall discuss using cassava starch in direct form, as hydrolysates, and as dextrins.

### ■ Starch in Direct Form

In Nigeria in the early 1990s, only about 700 tons of cassava starch was produced per year because Nigerian cassava starch is considered to be of low quality by Nigerian industries and none is exported. By contrast, in the early 1990s, more than 1 million tons of cassava starch was produced in Thailand, 40 percent of which was exported because the quality was high (Ratanawaraha, Senanarong, and Suriyapan 1999, 14).

In Nigeria, the bulk of the starch used as industrial raw material during the late 1990s, 17,000 tons, was corn starch, which represented 82 percent of the total. In the late 1990s, cassava starch accounted for only 600 tons, or 3 percent, of the 17,000 tons of starch used each year as industrial raw material in Nigeria. The remaining 15 percent was an unspecified type of starch. The quantity of industrial starch used in Nigeria in the late 1990s was low. Even if most of it were made from cassava, the impact on the total demand for cassava would be small. One of the reasons is that industrial production in Nigeria declined in the 1990s. For example, the amount of starch used in the various industries was significantly higher in the late 1980s (33,000 tons per year) than in late

1990s (17,000 tons per year) (Raw Materials Research and Development Council 2000b, 5).[5] The distribution of the total starch used in Nigeria was as follows: confectionary industry, 79.2 percent of total; automotive and dry cell battery, 6.1 percent; petroleum drilling, 5.5 percent; paint, 4.0 percent; textiles, 3.8 percent; and the remaining 1.4 percent in the pharmaceutical, iron ore mining, foundry, paper, soap and detergent, packaging, and cosmetic industries.[6]

*The Textile Industry.* In the textile industry, starch is used in the sizing operation, to coat yarn; in the finishing operation, to modify appearance, change stiffness, and add weight to fabric; and in the printing operation, to prepare the paste of dyestuff (Balagopalan et al. 1988). Nigeria's 198 spinning, weaving, or finishing textile plants produced 360 million meters of textile materials in 1992 (African Development Consulting Group 1997, 2). The textile mills use mostly imported corn starch. The director of the Nigerian Cotton Mill (NCM) in Onitsha recently reported that the NCM has discontinued the use of Nigerian cassava starch because it was of low quality.[7]

Cassava roots are often dried on the ground along the shoulders of highways, where they gather dust and mold. The starch industry should introduce a differential pricing system based on quality to induce smallholders to adopt better methods of drying cassava roots. There is potential for increased use of cassava starch as a raw material in the textile industry because Nigeria's large population has a strong demand for African prints. In 1992, Nigerian textile mills exported 30 percent of their total production, and the export was valued at US$130 million (African Development Consulting Group 1997).[8]

The Nigerian Starch Mill (NSM), Ihiala, the largest starch mill in Nigeria, produces cassava starch with dried roots purchased from smallholders.[9] In January 2001, the director of the NSM reported that he did not consider improving the quality of his product necessary because he was already able to sell all the starch he produced. Besides, if he invested in research and development to improve starch production technology, he would not have any patent protection.

*Petroleum Drilling Industry.* Starch is used as a clay dispersant in drilling mud in the petroleum industry (Balagopalan et al. 1988). There are two types of drilling mud: water-based and pseudo oil (synthetic fluid)-based drilling muds. The water-based mud is a mixture of water and starch, and it is used for drilling up to a depth of seven thousand feet. Beyond that depth, starch breaks down because of high temperature, and water-based mud is replaced with pseudo oil-based mud (Onwuta 2001). In Nigeria in early 2001, imported starch was being used in water-based drilling mud but other types of starch could be used if they gelatinize in cold water. The director of the Nigerian Starch Mill (NSM) was unaware that starch was used in oil drilling. He would like to sell starch to the oil industry but he is reluctant to invest in research to make cassava starch gelatinize in cold water because of the lack of patent protection.[10]

There is a large potential for the commercial use of tropical starches such as cassava starch in the petroleum industry, but considerable research and product development are necessary in order to achieve this potential (Satin n.d.). Since the oil sector supplies 95 percent of Nigeria's foreign exchange earnings, investment in research to make cassava starch acceptable to the petroleum drilling industry could lead to a large increase in the demand for cassava. Nigeria has an oil reserve of 22.5 billion barrels and it is currently producing 2 million barrels a day (MBendi 2000, 1 and 2).

## ■ Starch Derivatives: Hydrolysates

When starch is subjected to an acid (usually sulphuric acid) or enzyme treatment, sweet substances called starch hydrolysates are produced: glucose, sucrose, maltose, fructose, and syrup (Balagopalan et al. 1988, 147 and 148). Starch hydrolysates vary in sweetness and viscosity depending on the degree of the acid or enzyme treatment and dehydration. They are used to impart sweetness, texture, body, and cohesiveness to drinks such as soft drinks, fruit juices, and dairy drinks, and to a variety of foods, such as soup, cake, and cookie mixes; canned foods; and confectionaries. Starch hydrolysates are also a basic input in the manufacture of industrial chemicals such as alcohol, gluconic acid, and acetic acid (Balagopalan et al. 1988). In Nigeria, starch hydrolysates are used mostly in the soft drink and pharmaceutical industries.

*The Soft Drink Industry.* In Nigeria in the late 1990s, 174,000 tons of syrup concentrates were used in the soft drink industry to produce 33 million hectoliters of soft drinks per year (Raw Materials Research and Development Council 2000c, 23). The soft drink industry is dominated by Coca Cola, which imports the syrup concentrates and keeps them as a trade secret. Nigeria's soft drink industry imports all of its syrup concentrate because cassava starch is not currently hydrolyzed into syrup in Nigeria. In the early 1990s, however, an IITA postharvest technologist made syrup concentrate from cassava starch by treating it with sorghum enzyme. A pilot project is needed to test the suitability of cassava starch syrup concentrate in the preparation of soft drinks. If locally produced cassava starch could be converted into syrup concentrate and replace imported syrup, it would open up a market for almost 1 million tons of cassava per year, or about 3 percent of current cassava production.

*The Pharmaceutical Industry.* The current annual use of starch hydrolysates in the pharmaceutical industry is as follows: glucose, 771 tons; sucrose, 750 tons; and syrup, 2 tons.[11] At full capacity production, Nigeria's seventy-seven pharmaceutical plants could supply about 75 percent of Nigeria's current pharmaceutical needs. Yet Nigeria's plants are operating at only about 35 percent of their capacity because of a lack of basic raw materials (Raw Materials Research and Development Council 2000b, 13). Currently, 80 percent of the raw material used by the pharmaceutical industry is imported, because the basic chemicals and petrochemical industries are not developed in Nigeria (Raw Materials Research and Development Council 2000b, 12). The Raw Materials Research and Development Council (RMRDC) observed that "any reasons that can be advanced for the slow pace of industrialization in Nigeria can be reduced to one single fact: for now it is easier, less risky, and much less complicated to import than to manufacture locally" (RMRDC 1997, 41).

### ■ Starch Derivatives: Dextrins

Dextrins are produced by heating starch in a dry form with acid or alkali as a catalyst (Colonna, Buleon, and Mercier 1987, 110). Dextrins are sold as powders, granules, and pastes. Adhesives are made by cooking a dextrin in water. Adhesives are used in making corrugated boxes, sealing cartons, grocery bags, and multiwall bags in the packaging industry; for lamination in the plywood, paperboard, footwear, and cables industries; in the production of paper tubes, cans, and cones; as printing, publishing, and library paste; and as label adhesive for envelopes, postage stamps, gummed tapes, safety matches, and many other items.

In the late 1990s, an average of 58,000 tons of adhesives were used per year in the following industries in Nigeria: wood, 44 percent; cable, 26 percent; paper, printing, and publishing, 15 percent; packaging, 14 percent; and matches and footwear, 1 percent.[12] All the adhesives used in Nigeria in the late 1990s were imported either as adhesives or as dextrins. If all of the adhesives used in Nigeria had been derived from cassava, about 60,000 tons of cassava, representing 0.2 percent of Nigeria's current production, would have been utilized.

## Beer Malt: A New Use for Dried Cassava Roots

Beer has been brewed in Nigeria with imported barley malt for many decades. However, in 1985–86 Nigeria banned grain imports and the brewery industry began to produce beer malt with sorghum produced in northern Nigeria. The initial concern that sorghum beer would not be acceptable to consumers proved to be without basis, as beer consumption did not decline after sorghum malt was used to replace barley malt. In the late 1990s, about 11 million hectoliters of beer were being produced per year in Nigeria (Raw Materials Research and Development Council 2000c, 23).

Although cassava is produced in southern Nigeria, where all of the beer breweries are located, no attempt has been made to produce beer malt with dried cassava roots, even though dried roots are cheaper (US$79 per ton) than sorghum (US$139 per ton) (Ogazi, Hassan, and

Ogunwusi 1997, 31 and 77).[13] A biochemist of the National Root Crops Research Institute (NRCRI) at Umudike reported in early 2001 that beer malt could be made with any starch, provided the right type of enzymes are available.[14] Research is now needed to determine the type and quantity of enzymes needed for making beer malt from dried cassava roots. However, the NRCRI is unable to carry out the needed research because the operational budget for its research unit is only a few hundred dollars per year.

The manager of the Golden Guinea Brewery, Umuahia believes that consumers would accept cassava malt beer, judging from their ready acceptance of sorghum beer in the mid 1980s.[15] However, the manager reported that Golden Guinea would be reluctant to invest in research on making beer malt from cassava roots because patent law is not enforced in Nigeria.

In early 2001, the manager of the Life Beer Brewery in Onitsha reported that Life Beer is made directly from sorghum without malting at the rate of 9 tons of sorghum per 500 hectoliters of beer.[16] Using this ratio, the beer industry in Nigeria consumed about 200,000 tons of sorghum per year in the late 1990s. If dried cassava roots had replaced sorghum, the beer industry would have consumed 220,000 tons of dried cassava roots, which is more than 2 percent of annual cassava production in the 1990s. Research is needed on how to make beer malt from dried cassava roots, because even a partial substitution of dried cassava roots for sorghum in the beer malt will reduce the cost of beer production, increase employment in the beer industry, and raise the income of cassava farmers.

### Developing a Cassava-Based Alcohol/Ethanol Industry

In the late 1990s, 78 million liters of alcohol were used each year in Nigeria by the liquor industry and 10 million liters by the pharmaceutical industry.[17] Virtually the whole of the alcohol used in Nigeria in the late 1990s was imported, because domestic production was insignificant. For example, in 1998, the total production from Nigeria's sole ethanol plant was only 200,000 liters (Bamikole and Bokanga 2000).

In 1963, the Nigerian government set up a sugar plant, the Nigerian Sugar Company (NISUCO) to produce sugar from sugar cane. Ten years later, the government set up the Nigerian Yeast and Alcohol Manufacturing Company (NIYAMCO) as an annex to NISUCO with a goal of producing ethanol with molasses. Although NIYAMCO had an installed capacity of 4 million liters of ethanol per year, the supply of molasses began to decline in the early 1990s because of the collapse of the government-owned sugar plantation, which supplied sugar cane to NISUCO. In 1994, NIYAMCO began looking for an alternative source of raw material. With IITA's technical support, dried cassava root was selected as a raw material for the manufacture of ethanol by the NIYAMCO because cassava is abundant in Nigeria, has a high starch content, and has a low gelatinization temperature (Bamikole and Bokanga 2000). NIYAMCO required only about 30 tons of dried cassava roots per day but because of problems in organizing the collection of dried cassava roots from scattered smallholders, NIYAMCO closed its ethanol plant (Bamikole and Bokanga 2000). If the 88 million liters of alcohol currently imported each year for the liquor industry were produced with cassava roots in Nigeria, it would open up a market for about 600,000 tons of cassava roots, or about 2 percent of national cassava production during this period.[18]

In the United States, the Clean Air Act Amendment of 1990 mandated the sale of oxygenated fuels in areas with unhealthy levels of carbon monoxide. Since that time, there has been strong demand for ethanol as an oxygenate blended with gasoline. In the United States, more than 1.5 billion liters of ethanol is added to gasoline each year. Ethanol is blended with gasoline to form an E10 blend (10 percent ethanol and 90 percent gasoline) (Alternate Fuels Data Center 2000, 1). Major automobile manufacturers also have models that can operate on an E85 blend (85 percent ethanol and only 15 percent gasoline) (Alternate Fuels Data Center 2000, 3). Fuel ethanol blends are successfully used in all types of engines that require gasoline. However, ethanol production is subsidized in the United States.

The dynamic growth of Brazil's ethanol industry may awaken African policy makers and entrepreneurs to an opportunity for increased cassava production and utilization in cassava-based ethanol or alcohol production. The story of the Brazil's ethanol industry began in 1931, when the government enacted a decree that called for the addition of 5 percent ethanol to imported gasoline in order to find a market for Brazil's ailing sugar cane industry. Almost forty-five years later, in 1975, the government introduced the Brazilian National Alcohol Program to reduce the imported crude oil bill by increasing the percentage of ethanol in gasoline from 5 percent to 20 percent (Lindeman and Rocchiccioli 1979; Pimentel 1980). To realize this goal, the Brazilian government set up two research and development agencies, the Technological Alcohol Program and the Sugar and Alcohol Institute.

The new agencies explored the sources of various raw material and found that of the sources available, "the most important by a wide margin was cassava root. . . . It was evident that cassava was a very viable alternate source for ethanol" (Lindeman and Rocchiccioli 1979, 1108 and 1109). The yield of alcohol per ton is higher from cassava (150 liters per ton of fresh root) than from sugar cane (48 liters per ton) (Balagopalan et al. 1988, 182). A much lower quality soil is required to produce cassava than sugar cane, and a distillery using cassava to produce alcohol can operate year-round because cassava is available year-round. By contrast, a distillery based on sugar cane as a raw material must stand idle for several months a year when sugar cane is not available (Lindeman and Rocchiccioli 1979).

Brazil's research and development agencies compared sugar cane and cassava and selected sugar cane to produce ethanol because sugar cane was an old and traditional crop that was cultivated on a large scale throughout the country. Sugar cane varieties in Brazil were nearly all hybrids that had been developed by the national experimental stations, except for a few obtained from India and Argentina (Pimentel 1980, 2002).

Turning from production to processing sugar cane, research in Brazil showed that there was a great potential for small-scale alcohol processing units for developing countries (Balagopalan et al. 1988). However, Brazil opted for large units, because sugar cane, the chosen raw material source, is produced on large plantations. Another challenge was what to do with the expected large amounts of ethanol and sugar industrial wastes such as stillage and bagasse. The research agencies recommended recycling stillage as fertilizer for sugar cane production and using bagasse to generate electricity for the ethanol plants (Pimentel 1980 and GreenTimes 2000).[19]

Brazil's learning-by-doing approach yielded positive results! Ethanol production grew from a mere 0.6 billion liters in 1974–75 to 14.0 billion liters in 1998–99 (Lindeman and Rocchiccioli 1979, 1111; GreenTimes 2000, 1). Brazil was able to achieve this level of success in ethanol production because large-scale sugar cane producers responded positively to the expanding market for ethanol: the area of sugar cane harvested increased from 1.9 million hectares per year in the early 1970s to 4.9 million hectares per year in the late 1990s; yield increased from an average of 46.7 tons per hectare in the early 1970s to an average of 68.1 tons per hectare in the late 1990s; and total production increased from 89 million tons per year in the early 1970s to 331 million tons per year in the late 1990s (FAOSTAT). In the year 2000, one million people were working in Brazil's sugar-ethanol sector, including three hundred thousand in 350 private ethanol units and fifty thousand sugar cane growers (GreenTimes 2000, 1).

Based on research in Brazil, Nigeria and other cassava-producing African countries may be able to produce ethanol or alcohol with small-scale cassava-based production units (Balagopalan et al. 1988). Although Nigeria could theoretically benefit by using cassava to produce alcohol and replace alcohol imports for alcoholic beverages, public enterprises such as the NIYAMCO and NISUCO have floundered in Nigeria because of mismanagement of public resources and the inability of the government to provide research and development support to assist new industrial enterprises. However, a cost-benefit study of ethanol production should be completed in petroleum-importing Ghana. A cassava-based ethanol industry could reduce the petroleum import bill for major cassava-producing and petroleum-importing African countries such as the Congo, the Côte d'Ivoire, Ghana, Tanzania, and Uganda.

## Summary

Research is urgently needed on developing new uses for cassava in livestock feed and as industrial raw material. We have pointed out that Thailand dominates world trade in cassava pellets for the livestock industry in the European Union. Yet with increasing competition from U.S. grain, the price of pellets has fallen and Thailand's exports have declined from 9 million tons in 1989 to 3 million tons in 1998. Faced with depressed world prices of cassava pellets, African cassava-producing nations should give priority to expanding the use of cassava in livestock feed in Africa rather than to exporting cassava pellets to the European Union.

What can be done to increase the use of cassava in livestock feed in Nigeria? Currently, only 5 percent of Nigeria's cassava production is used in livestock feed, as compared with 50 percent in Brazil, because in Nigeria, cattle, sheep, and goats are free grazed and pigs rummage on household waste. Also, the poultry industry is small in Nigeria, with only 125 million birds, as compared with 867 million birds in Brazil in 1998. Yet a feeding trial in Nigeria has shown that a combination of cassava roots and leaves in a ratio of four to one will allow cassava to replace maize in poultry feed. If this feed ratio is adopted by farmers, it could lead to a significant increase in the quantity of cassava roots used in poultry feed.

There are technologies for use of cassava as a partial substitute for wheat in food manufacture. But in Africa, cassava is not used for food manufacture because food products made with 100 percent wheat flour are cheaper and preferred by consumers. Sustained investment in research and development on industrial manufacture of African cassava food products such as *gari, attieke,* and *chickwangue,* which have snack values, can lead to increased use of cassava in food manufacturing industries.

Turning to industrial uses for cassava, the scope for increasing the use of cassava starch in Nigeria's industries is, to a large extent, determined by the ready availability of high-quality imported starch and Nigeria's meager research and development investment in preparing cassava starch for industrial uses. Many of Nigeria's industrial products are not competitive

**Table 12.1. Nigeria: Industrial Potential for Cassava and Research and Development Priorities.**

| INDUSTRY | CASSAVA-BASED RAW MATERIAL | POTENTIAL FOR INCREASING CASSAVA USE | REQUIRED RESEARCH AND DEVELOPMENT ACTIVITY |
|---|---|---|---|
| Textile | Cassava starch in direct form | Medium | Improve the quality of cassava starch by improving the method of drying cassava roots |
| Petroleum drilling | Cassava starch in direct form | Medium | Make cassava starch gelatinize in cold water |
| Pharmaceutical | Cassava starch hydrolysates including glucose, maltose, sucrose, fructose, and syrup | Medium | Set up industries to make glucose, maltose, sucrose, fructose, and syrup from cassava starch |
| Soft drinks | Syrup concentrate | High | Prepare syrup concentrate from cassava starch and test it for suitability for making soft drinks |
| Beer | Dried cassava roots | High | Develop a method of making beer malt using dried cassava roots |
| Ethanol | Dried cassava roots | High | Set up cassava-based small-scale alcohol units; carry out feasibility study of producing ethanol from cassava in Nigeria and Ghana |

with imported goods in terms of quality and cost. In addition, inadequate transport, power supply, security, research and development, and manpower training have made industrial production in Nigeria inefficient and uncompetitive globally. The Nigerian industries that use starch as a raw material have access to cheaper and higher quality starch from imported sources. In addition, Nigerian private investors do not have the incentive to invest in cassava research and development because of the lack of patent protection.

Table 12.1 illustrates the potential industrial uses for cassava, and research priorities for developing cassava starch and dried cassava roots as industrial raw materials in six different industries in Nigeria: the textile, petroleum drilling, pharmaceutical, soft drink, beer malt, and ethanol/alcohol industries. The potential is high in three industries. The first is syrup concentrate for the soft drink industry in Nigeria. Syrup concentrate has been successfully made from cassava starch by an IITA postharvest technologist. A pilot project is needed to determine its acceptability and potential profitability in making soft drinks.

The second potential use of cassava is in the beer industry. The Nigerian beer industry currently uses about 200,000 tons of sorghum each year to make beer malt. No attempt has yet been made to prepare beer malt from dried cassava roots. However, biochemists at the National Root Crops Research Institute (NRCRI) believe that given the right enzyme, it is possible to prepare beer malt from dried cassava roots. Research is needed to develop the technology for making beer malt from dried cassava roots.

The third potential for cassava as an industrial raw material is a cassava-based alcohol industry. Currently, Nigeria imports about 90 million liters of alcohol annually, with about 80 million liters being used by the liquor industry. If the 80 million liters were produced from cassava, it would require 500,000 tons of dried cassava roots, which would increase the demand for cassava, raise farm income, generate on-farm and off-farm jobs, and save foreign exchange.

Yet a cassava-based ethanol industry should be of more help to other cassava-producing countries such as Ghana than to Nigeria, because Nigeria is subsidizing the retail price of gasoline while Ghana is importing petroleum. A feasibility study should be carried out to determine the economics of a cassava-based alcohol industry in Nigeria and a cassava-based ethanol industry in Ghana.

CHAPTER 13

# The Cassava Transformation: Synthesis

## Cassava and Africa's Food Crisis

Africa's food crisis is a stubborn problem. Africa became a net food importer in the early 1970s, and food production grew at half of the population growth rate from 1970 to 1985. Africa's population is expected to double to 1.2 billion by 2020, and its urban population will grow at a faster rate. The average per capita GNP in Africa in the year 2000 was US$480 (World Bank 2000). Many countries have been destabilized by civil strife, authoritarian regimes, and the HIV/AIDS pandemic. Yet the African food crisis is rooted in the mass poverty of Africans and the low and unstable crop yields throughout the nation. Raising per capita incomes and increasing food production represent two sides of the same coin (Eicher 1982).

The dramatic cassava transformation that is under way in some African countries such as Nigeria and Ghana is Africa's best-kept secret. This book describes how the new TMS varieties have transformed cassava from a low-yielding famine-reserve crop to a high-yielding cash crop that is prepared and consumed as a dry cereal (*gari*). We focus on the cassava transformation for two important reasons. First, cassava is Africa's second-most important food staple in terms of per capita calories consumed. Cassava is a major source of calories for roughly two out of every five Africans. In some countries, cassava is consumed daily, and sometimes more than once a day. In the Congo, cassava contributes more than one thousand calories per person per day to the average diet, and many families eat cassava for breakfast, lunch, and dinner. Cassava is consumed with a sauce made with ingredients rich in protein, vitamins, and minerals. In the Congo and Tanzania, cassava leaves are consumed as a vegetable. Cassava leaves are rich in protein, vitamins, and minerals.

The second reason to focus on cassava is to tell the story of the cassava production revolution that has been fueled by the introduction of the high-yielding TMS varieties, starting in Nigeria in 1977. The TMS varieties have boosted farm-level yields by 40 percent (from 13.41 tons to 19.44 tons per hectare) without fertilizer application.

With the aid of mechanical graters to prepare *gari,* cassava is increasingly being produced and processed as a cash crop for urban consumption in Nigeria and Ghana. This book points out the potential of cassava in helping to close Africa's food production gap and helping the rural and urban poor by reducing the price of cassava. Yet the underlying data base on cassava in Africa is woefully inadequate. Only two books have been published on cassava in Africa in the past forty years: *Manioc in Africa* by W. O. Jones (1959) and *Cassava in Shifting Cultivation* by L. O. Fresco (1986).

To address the lack of information on Africa's second-most important food crop, a Collaborative Study of Cassava in Africa (COSCA) was initiated by the Rockefeller Foundation in 1989 and based at the International Institute of Tropical Agriculture (IITA). During an eight-year period (from 1989 to 1997), COSCA researchers collected information from 1,686 households in 281 villages in six countries: the Congo, the Côte d'Ivoire, Ghana, Nigeria, Tanzania, and Uganda. In early 2001, COSCA researchers collected information from cassava starch industries and from industries that use cassava as a raw material in Nigeria. The findings of the six-country COSCA study are the primary source of information for this book.

Without question, cassava is a much-maligned crop. Many critics of cassava have pointed out that:

- cassava is consumed by poor rural households and by animals;
- cassava impoverishes the soil;
- cassava is a "women's crop" that requires female extension agents and special cassava projects for women;
- cassava often kills people because it contains lethal quantities of cyanogens (prussic acid); and
- cassava is deficient in protein.

These myths and half-truths constitute a great deal of misinformation which has discouraged African governments and donors from investing in speeding up the cassava transformation. The COSCA findings presented in this book can help African policy makers and scientists update their overall knowledge about cassava and its role as an important component of anti-poverty and food security programs.

Cassava is a food staple in low-, middle-, and upper-income households in both rural and urban areas. The COSCA data show that one-third to two-thirds of the cassava planted in Africa is destined for urban markets. The COSCA soil fertility studies show that cassava fields, some of which have been under continuous cultivation for at least ten years, are as fertile as fields planted to other crops. The contention that cassava is a "women's crop" should be viewed as an important half-truth because the COSCA studies reveal that cassava is both a "women's crop" and a "men's crop." In most COSCA study countries, men own a higher percentage of cassava fields than women.

The assertion that some cassava varieties contain cyanogens (prussic acid), which are sometimes lethal, is also a half-truth because the cyanogens can be eliminated by using well-known traditional processing methods. The fear of cyanogens should not discourage public or private investment in the cassava food economy. We agree with nutritionists that cassava is deficient in protein and vitamins. However, instead of discouraging cassava production and consumption because the plant is low in protein, policy makers should focus on helping smallholders diversify their crops and increase rural incomes so that families can purchase protein-rich foods to supplement their diets.

Cassava is the cheapest source of food calories in most countries where it is widely grown. Most of the scientific research and donor attention in Africa has focused on cassava's role as a food crop. Yet the heavy emphasis on cassava as a food crop is selling cassava short! Cassava can play the following roles in African development, depending upon the stage of the cassava transformation in a particular country:

- famine reserve,
- rural food staple,
- cash crop for urban consumers,
- industrial raw material, or
- livestock feed.

The first three roles currently account for about 95 percent of Africa's cassava production, while the last two account for 5 percent. Cassava performs a famine-reserve function in countries such as Tanzania, where rainfall is too unreliable to assure a stable maize supply. Cassava plays a rural food staple role in the forest zone in some countries, such as the Congo. In Nigeria and Ghana, cassava is playing an increasingly important role as a cash crop for urban markets. With access to improved processing methods, farmers are able to prepare an array of cassava products for urban consumers. The challenge is to intensify public and private sector research and development on new uses for cassava that can enhance cassava's contribution to development.

In the 1960s, Brazil was the world's leading cassava producer and Africa accounted for only 40 percent of world production. Thirty years later, in the 1990s, Africa produced half of the world cassava output and Nigeria had replaced Brazil as the leading cassava-producing country in the world. The factors responsible for the dramatic expansion of cassava production in Africa are:

- the introduction of the mechanized grater and *gari* preparation methods that have transformed cassava into a dry cereal for urban consumption.

- rapid population growth and poverty, which have encouraged consumers to search for a cheaper source of calories
- high-yielding TMS varieties that have boosted farm-level cassava yields by 40 percent in Nigeria and Ghana, and
- biological control of the cassava mealybug

Although population pressure on land is increasing, the TMS varieties have increased cassava yields in Africa. The COSCA researchers studied the relationship between technological change and population densities in three villages comprising low, medium, and high population densities in southeastern Nigeria. Over the 1973 to 1993 period, cassava yields doubled in the high population density village because of the introduction of the TMS varieties.

### Genetic Research and the High-Yielding TMS Varieties

The evolution of cassava breeding in Africa can be described as a human ladder (see table 5.4). Two British colonial researchers, H. H. Storey and R. F. W. Nichols, launched cassava research at the Amani research station in Tanzania in 1935. They developed Ceara rubber x cassava hybrids that were resistant to the mosaic and brown streak diseases. D. L. Jennings took over the cassava research program at the Amani research station in 1951 and developed Ceara rubber x cassava hybrids that had a wider resistance to the diseases. As African countries regained independence in the late 1950s, cassava research languished, and the Amani research program was terminated in 1957.

Dr. S. K. Hahn joined the International Institute of Tropical Agriculture (IITA) in 1971 and initiated a cassava crossbreeding program by drawing on the germplasm of the Amani researchers and adding earliness of bulking, high yield, good root quality, and resistance to lodging. After only six years of research, Hahn and his IITA team hit the jackpot in 1977 when they released the high-yielding TMS cassava varieties. The TMS varieties increased farm-level yields by 40 percent without fertilizer and were superior to local varieties in terms of earliness of bulking and pest and disease tolerance. Dr Hahn retired from the IITA in 1994, after twenty-three years (from 1971 to 1994) of outstanding leadership. Dr A. G. O. Dixon assumed the leadership of IITA's cassava improvement program. The IITA's current cassava breeding goals are to enhance earliness of bulking and the ease of peeling, to develop appropriate canopy sizes for harvesting leaves for consumption and intercropping, and to improve resistance to pests and diseases. Although the mealybug has been brought under control through the biological control program of the IITA, green mite, bacterial blight, and mosaic disease are still serious constraints on cassava production in various parts of Africa.

COSCA researchers found that commercial cassava farmers in Nigeria and Ghana desire varieties that will attain a maximum yield in fewer than twelve months so that they can grow cassava on the same field for a number of years in succession. Yet farmers in Nigeria who plant the TMS varieties and prepare *gari* for sale in urban centers are facing serious labor bottlenecks at the harvesting and peeling stages. Research is urgently needed on reducing the bulking period and developing mechanized harvesters and peelers.

## The Diffusion of the TMS Varieties

### ■ Rapid Diffusion in Nigeria

The diffusion of the TMS varieties in the late 1970s and the 1980s in Nigeria was facilitated by the unflagging leadership of Dr S. K. Hahn, the head of the IITA's cassava program. Under his leadership, the IITA's cassava diffusion program multiplied and distributed TMS planting materials to farmers both directly and indirectly, through informal channels such as schools and churches. The IITA's diffusion program mobilized the private sector, particularly the petroleum industry, to assist them in the testing and distribution of the TMS varieties. The IITA's cassava program also mobilized the mass media, including newspapers, radio, and television, to publicize the availability and benefits of the TMS varieties. By the late 1980s, the TMS varieties were grown in 60 percent of the villages in the cassava-growing areas of Nigeria. The TMS 30572 variety was the most popular among farmers. It was widely planted by farmers who used it to prepare *gari* for urban markets.

The diffusion of the TMS varieties was delayed in other COSCA study countries, however, because their governments did not put in place the institutions and financial resources for on-farm testing, multiplication and distribution that Nigeria's government had.

### ■ Delayed Diffusion in Ghana

The biggest question about the cassava transformation in Africa is why the high-yielding TMS varieties were not released in Ghana until 1993, some sixteen years after they were released in Nigeria in 1977. This is a puzzle because the TMS varieties increased farm-level yields by 40 percent, which is roughly the same as the increases in rice and wheat yields during Asia's Green Revolution in the 1960s. Yet while high-yielding rice and wheat varieties were quickly diffused throughout Asia in the 1960s, there was a long delay before farmers in Ghana had access to the new cassava varieties.

The answer to the TMS diffusion puzzle in Ghana has now been pieced together. Ghana's official release of the TMS varieties in 1993, sixteen years after they were released in Nigeria, is attributed to a variety of factors. First, the government had historically invested in research on maize, the primary food staple, and initiated cassava research in its National Agricultural Research System only in 1980. Second, the government of Ghana displayed little interest in cassava in its food policies in the 1970s and 1980s. Third, it took five years, from 1988 to 1993, for government researchers to evaluate the TMS varieties under field conditions and multiply the stem cuttings.

To summarize, Ghana's TMS diffusion experience can be described as a delayed success story. Unfortunately, however, the TMS diffusion in Ghana since 1993 has not been documented in terms of the area planted. In early 2001, researchers at the Crops Research Institute in Kumasi reported that the TMS varieties covered large areas of farmers' fields in the Eastern, Greater Accra, and Volta regions, where farmers were preparing *gari* for sale in urban centers. The most common TMS varieties grown by farmers in those regions is Afisiafi (TMS 30572) (Otoo and Afuakwa 2001).

### ■ Uganda's Experience

In Uganda, government and donor interest in the diffusion of the TMS varieties was sparked by the appearance and rapid spread of cassava mosaic disease in the late 1980s. From 1991 to 1996, the National Agricultural Research Organization collaborated with the National Network of Cassava Workers, Non-Governmental Organizations (NGOs), and district and local leaders to conduct accelerated on-farm trials, training of farmers, and multiplication and distribution of mosaic-resistant varieties, namely TMS 6014, TMS 30337, and TMS 30572. Over 35,000 farmers, opinion leaders, and extension agents were trained and 536 on-farm trials were conducted (World Bank 2001). By the year 2000, about 80,000 hectares of the mosaic-resistant TMS varieties were under cultivation (Otim-Nape et al. 2000).

## Postharvest Issues: Food Preparation and Mechanized Processing

There are five common cassava food products in Africa: fresh cassava roots, dried cassava root flour, pasty cassava products, granulated cassava products, and cassava leaves. The traditional methods used to make the products have been influenced by the availability of sunlight, water, fuel wood, and market demand. Cassava food preparation methods vary widely in Africa. The consumption of *gari* is common throughout West Africa but not in Central and East Africa. The consumption of cassava leaves is common in the Congo and Tanzania but not in the Côte d'Ivoire, Ghana, Nigeria, or Uganda. The challenge is to promote the diffusion of the best practices in cassava preparation throughout cassava-producing economies in Africa, with an aim of improving the economic and nutritional welfare of all cassava producers and consumers.

There are three types of cassava-processing machines in use in Africa: graters, pressers, and mills. The traditional method of grating cassava by pounding it in a mortar with a pestle was replaced by the use of the manual grater, which was made out of a sheet of perforated metal mounted on a flat piece of wood. Mechanized graters were first introduced in the Benin Republic by the French in the 1930s and were modified in Nigeria in 1940s by welders and mechanics, using local materials such as old automobile motors and scrap metal. The mechanized graters are owned by village entrepreneurs, who provide grating services to farmers. The mechanized grater operators allow the farmers flexibility in terms of working time and quantity of cassava grated. The mechanized grater is rapidly replacing manual graters because it reduces processing labor by 50 percent and dramatically increases the profitability of *gari* production.

It is important to point out, however, that *gari* preparation technology (converting cassava into dry cereal) and the mechanized grater were in use in Nigeria before the TMS varieties were released to farmers in the late 1970s. The ready availability of mechanized processing equipment and *gari* preparation technology increased the diffusion of the TMS varieties, because mechanized grating is cheaper than hand grating, and it reduces the need to recruit and manage a large labor force. The Nigerian/Ghanaian mechanized grater that is made with local materials should be introduced in other cassava-producing countries because it is reliable and cost-effective. Since the grating task has been mechanized, peeling is now the most labor-intensive task in *gari* preparation, followed by the toasting stage. There is a need for engineers and breeders to join forces to develop cassava roots that are uniform in shape and size in order that they can be peeled with a machine. There is also a need to develop methods that will reduce the amount of toasting labor in *gari* preparation.

Private entrepreneurs have made many attempts to process cassava on a large scale using integrated grating, pressing, pulverizing, sieving, and toasting machines in a factory-style operation. However, the cassava-processing factories in the Congo, the Côte d'Ivoire, Ghana, and Nigeria have all failed because of a shortage of cassava roots; a fragmented market for cassava products; a high cost of peeling by hand, and quality control problems. An array of government agencies were established in Nigeria in the 1970s and 1980s to develop mechanized cassava-processing machines. Yet the machines developed by these agencies have not been adopted by farmers because they are not as convenient and reliable as those developed by the small-scale private artisans.

### Gender Surprises

One of the myths of cassava is that it is a "women's crop" that requires special assistance from female extension agents. We tested this widely held belief by collecting data from 1,686 rural households in 281 villages in the six COSCA study countries. Five gender surprises emerged from the COSCA surveys:

- Although men and women specialize in different tasks in the cassava food system, both contribute a significant amount of their labor to the cassava sector.
- Women contribute less than half of the total labor involved in the production and processing of cassava in all of the COSCA countries except the Congo.
- As the cassava transformation proceeds and cassava becomes a cash crop produced primarily for urban centers, men increase their labor contribution to each of the production and processing tasks.
- The introduction of labor-saving technologies in cassava production and processing has led to a change in gender roles. For example, in Nigeria, when *gari* was processed by manual grating, it was done by women. Yet when the grating process was mechanized, it was taken over by men, who own and operate the mechanized graters.
- Women have been able to secure access to improved cassava production technologies and essential inputs such as land and male labor in all six COSCA study countries.

One must look beneath the surface of these macro findings, however, and examine country-specific situations. With the introduction of the TMS varieties, cassava has been transformed into a profitable cash crop in Nigeria and Ghana, and it has induced men to remain in the villages and produce cassava and process it into *gari* for sale to urban consumers. Yet in the Congo, young men have migrated from the villages to escape rural poverty, and women are therefore left to grow, harvest, and process cassava for their families under conditions of extreme rural poverty.

There is a need for NGOs to help diffuse improved cassava food preparation methods for women because they are solely responsible for this activity. For example, because of the surprising lack of knowledge about *gari* preparation in Central and East Africa and about the nutritional value of cassava leaves in the Côte d'Ivoire, Ghana, Nigeria, and Uganda, special educational programs are needed to help women gain access to the best practices in food preparation in the other major cassava-producing countries in Africa. There is also an urgent need for NGOs to help develop improved mechanical peelers, because women bear the brunt of the high-labor intensity of this task.

### New Uses for Cassava

Cassava in Africa is used almost exclusively for food. In fact, 95 percent of the total cassava production, after accounting for waste, is used as food. By contrast, 55 percent of total production in Asia and 40 percent in South America are used as food. In Africa, total cassava consumption has more than doubled, from 24 million tons per year in the early 1960s to 58 million tons per year in the late 1990s. The large increase in the total cassava consumption in Africa is due to a significant increase in per capita consumption in countries such as Nigeria and Ghana, where cassava is produced mainly as a cash crop for urban consumption.

Many international agencies and bilateral donors are hesitant to extend loans and grants to African nations to help them increase the production of root crops such as cassava because of the long-held belief that cassava is an "inferior good"—that is, cassava is consumed only by the poor and by animals. Yet the COSCA study found that the income elasticities of demand for cassava products among rural households were positive at all income levels. Also, in Nigeria and Tanzania, the income elasticity of demand for cassava roots consumed as vegetable in rural households was above one. In Nigeria, the income elasticity of demand for *gari* was significantly higher than that of maize at all income levels among rural households. In Nigeria and Ghana, the income elasticity of demand for cassava was higher among urban than among rural households. These robust income elasticities of demand show that there is a strong market demand for cassava products, and that the view of cassava as an inferior food is invalid.

Cassava was found to be the cheapest source of calories among all food crops in each of the six COSCA study countries. As family incomes increased, the consumption of cassava dried root flour declined, while the consumption of cassava in convenient food forms such as *gari* increased. Dried cassava root flour is cheaper than *gari* because of the high cost of processing *gari*. Many medium- and high-income families consume *gari* because it is cheaper and requires less cooking time than grain. The future of cassava as a rural and urban food staple will depend upon the cassava's ability to compete with wheat, rice, and maize in terms of cost, convenience, and availability in urban markets.

However, promoting cassava merely as food is selling cassava short! The cassava transformation critically depends upon finding additional markets for the increased production resulting from the success of the TMS varieties. In Nigeria, only 5 percent of cassava production is used as livestock feed because the livestock industry is small and underdeveloped. Cassava roots are used in poultry feed, but a poultry feed trial has shown that cassava roots and leaves combined in a ratio of four to one could replace maize in poultry feed and reduce feed cost without a loss in weight gain or egg production. This research finding should be diffused to farmers and livestock feed producers.

There are technologies for use of cassava as a partial substitute for wheat in food manufacture. But in Africa, cassava is not used for food manufacture because food products made with 100 percent wheat flour are cheaper and preferred by consumers. Sustained investment in research and development on industrial manufacture of African cassava food products such as *gari, attieke,* and *chickwangue,* which have snack value, can lead to increased use of cassava in food manufacturing industries.

In Nigeria, an insignificant proportion of cassava production is used as starch for industrial raw material, because industrial production is inefficient and Nigerian manufacturers have easy access to cheaper and higher-quality imported starch. Research on cassava starch is needed to improve its quality in order to meet specific industry needs.

In Nigeria, the soft drink industry imports about 170,000 tons of syrup concentrate per year. IITA scientists have successfully made syrup concentrate from cassava starch. A pilot project needs to be carried out to determine the feasibility of making syrup concentrate with cassava starch for the Nigerian soft drink industry.

Industry experts believe that dried cassava roots can be used to make malt for the brewing of beer. They also believe that consumers would accept beer made from cassava malt, just as they did when barley imports were banned in 1985 and beer was made from sorghum malt. Research and development are needed to develop technology for making beer malt with dried cassava roots.

Turning to using cassava to produce alcohol or ethanol, Brazil has been a pioneer in producing ethanol from sugar cane and blending ethanol with gasoline to reduce petroleum imports. Because of a long legacy as a sugar cane producer, Brazil opted to produce ethanol from sugar cane rather than from cassava. Brazil currently requires all gasoline to contain 20 percent ethanol. Brazilian scientists have also carried out numerous studies of small-scale ethanol plants. These plants should be studied by cassava-producing African countries that import petroleum. If cassava-based ethanol production turns out to be feasible in Africa, smallholders will have opened up a new market for cassava. Yet an economic study is needed to determine the profitability of a cassava-based ethanol industry and the feasibility of blending ethanol with gasoline in petroleum-importing countries such as Ghana and the profitability of cassava-based alcohol industry in petroleum-exporting countries such as Nigeria.

## Policy Implications

### ■ Cassava Can Be a Powerful Poverty Fighter

Increasing the productivity of the cassava food system can help both the rural and urban poor by driving down real (inflation-adjusted) cassava prices over time. However, since Africa's independence in the 1960s, cassava production has been promoted by an array of special projects instead of through a systematic attack on improving the productivity of the entire cassava food system. It is now time to focus sustained African and donor attention on the entire cassava production and utilization system and to develop action plans to accelerate the cassava transformation. The solution to poverty and low returns to the men and women engaged in cassava production and processing is to concentrate on introducing improved varieties, agronomic practices, and labor-saving harvesting and processing technologies, and finding new industrial uses for cassava. There is a need to remove subsidies on imported rice and wheat in order to provide a level playing field for cassava in the food system.

### ■ Breaking New Bottlenecks During the Cassava Transformation

Without question, increased cassava production is a powerful but incomplete engine of rural economic growth. The fuel that drives the cassava transformation is the introduction of biological and mechanical technologies coupled with institutional innovations and balanced food policies. The introduction of new technology creates new bottlenecks that have to be broken by further innovations and changes in policy. Several examples illustrate this substitution process at work. The spread of the TMS varieties in Nigeria has shifted the labor constraint from the production to the harvesting task. The biological control of the mealybug shifted the control program to the green spider mite and to the Ugandan variant of the mosaic disease. Likewise, the substitution of mechanized for hand grating has shifted the labor constraint to the peeling task. The lack of knowledge of *gari* preparation technology has restricted its use in the Congo, Tanzania, and Uganda. Finally, the lack of knowledge about how to prepare cassava leaves has prevented consumers in the Côte d'Ivoire, Ghana, Nigeria, and Uganda from taking advantage of the nutritional value of the leaves in their diets.

The TMS story illustrates the substitution process at work. The TMS varieties attain a maximum yield in fifteen months after planting, but commercial farmers in Nigeria are demanding varieties that will achieve a maximum yield in twelve months or less, so that they can practice continuous cultivation of cassava. To address the new bottlenecks that emerge during the cassava transformation, research and diffusion should be broadened beyond plant breeding and protection to include the following:

- development and diffusion of labor-saving technologies for harvesting cassava;
- development and diffusion of labor-saving methods for peeling cassava;
- development of the "market pull" for cassava by improving roads;
- development of industrial uses for cassava;
- diffusion of labor-saving cassava grating technologies to regions and countries where such technologies do not exist; and
- diffusion of knowledge of *gari* preparation methods to regions and countries where such methods do not exist;
- diffusion of knowledge of methods for preparing cassava leaves in regions and countries where such knowledge does not exist;

The exploratory survey of industrial uses for cassava in Nigeria that was undertaken by COSCA researchers in early 2001 and reported here in chapter 12 should be supplemented with an in-depth study of industrial uses for cassava in the major cassava-producing countries in Africa. Cassava farmers in Africa should be connected to the Internet to allow them access to market information worldwide. The information technology is now simple enough for this purpose. For example, wireless communication systems and solar electric generators are now available at a reasonable cost in some African countries such as Ghana and Nigeria.

## ■ The Need for Country-Specific Cassava Transformation Strategies

Country-specific cassava transformation strategies are needed because the role of cassava varies significantly among African countries depending upon the stage at which each country finds itself in the cassava transformation, cultural factors, agro-ecology, and market opportunities. This variation will determine whether cassava improvement strategies should emphasize high-yielding and early bulking varieties, labor-saving technology for harvesting, food preparation methods, labor-saving methods for grating, or the use of cassava in livestock feed and as industrial raw material.

Table 13.1 identifies challenges for the six COSCA study countries that are consistent with the stage of the cassava transformation in which each country finds itself. In both Nigeria and Ghana, the development of labor-saving mechanical harvesters and peelers will help to make cassava more competitive with food grains in urban markets and with other plant starches that are used in livestock feed and as raw materials in industries. Also, cassava varieties that can attain a maximum yield in fewer than twelve months will enable commercial farmers to grow cassava under continuous cultivation. To utilize the expanded production from the TMS varieties and improved agronomic practices, both Nigeria and Ghana need to more aggressively pursue new uses for cassava. International donor agencies should finance projects to develop new cassava food products and industrial uses, such as preparing dried cassava roots as malt for beer brewing, developing cassava syrup concentrate for soft drinks, encouraging small-scale cassava-based alcohol or ethanol production, and manufacturing and utilizing cassava starch.

In Uganda, research, government, NGO and donor agencies are struggling to control the devastating cassava mosaic disease by introducing the TMS varieties. TMS varieties are reported to be spreading, and farmers are getting higher yields (World Bank 2001). Government, NGO, and international donor agencies in Uganda should continue to address disease control. The diffusion of *gari* preparation methods and the mechanized grater should be promoted because they can reduce processing costs and expand the urban demand for cassava. NGO and donor agencies should support a

**Table 13.1. The Cassava Transformation: Country Challenges.**

| COUNTRY | POLICY CHALLENGES | GENETIC RESEARCH, PLANT PROTECTION, AGRONOMIC PRACTICES, TMS DIFFUSION | FOOD PREPARATION AND PROCESSING |
| --- | --- | --- | --- |
| Nigeria | • Research and development support for increasing the use of cassava in industry | • Develop varieties that attain a maximum yield in less than 12 months<br>• Develop labor-saving technology for cassava harvesting | • Develop labor-saving method for toasting *gari*<br>• Diffuse the cassava leaves preparation method |
| Ghana | • Research and development support for increasing the use of cassava in industry | • Develop varieties that attain maximum yield in less than 12 months<br>• TMS multiplication and diffusion<br>• Develop labor-saving technology for cassava harvesting | • Develop labor-saving method for toasting *gari*<br>• Diffuse the cassava leaves preparation method |
| Uganda | • Research and development support for increasing the use of cassava in industry | • TMS multiplication and diffusion<br>• Develop labor-saving technology for cassava harvesting | • Diffuse *gari* preparation methods<br>• Diffuse mechanized grater technology<br>• Diffuse the cassava leaves preparation method |
| Côte d'Ivoire | • Eliminate subsidies on imported rice and wheat | • TMS multiplication and diffusion<br>• Develop labor-saving technology for cassava harvesting | • Develop a grater that is suitable for *attieke* processing<br>• Diffuse the cassava leaves preparation method |
| Tanzania | • Improve road access to urban market centers | | • Diffuse *gari* preparation methods<br>• Diffuse the mechanized grater technology |
| Congo | • Improve civil stability<br>• Road access to urban market centers<br>• Eliminate import subsidies on rice and wheat | | • Diffuse *gari* preparation methods<br>• Diffuse the mechanized grater technology |

pilot project on using cassava as a raw material in beer malt, syrup concentrate for soft drinks, and the ethanol industry, as well as development of cassava starch as a raw material for other industries.

In the Côte d'Ivoire, the "market pull" of cassava can be increased by eliminating the subsidies on imported rice and wheat. A grater for making *attieke* needs to be developed to reduce the cost of a product that has a strong urban demand. In Tanzania, maize is the preferred staple and cassava is a famine-reserve crop. The diffusion of the *gari* preparation technology and the mechanized grater can help cassava compete in the urban market. International donor agencies should support pilot activities to find nonfood uses for cassava. In the Congo, increasing "market pull" by improving roads and removing subsidies on imported rice and wheat will help cassava farmers tap new rural and urban markets.

In summary, in Africa cassava is being transformed from a famine-reserve crop to a cash crop for urban consumption. The fuel that drives the cassava transformation is the new TMS varieties that have changed cassava from a low-yielding famine-reserve crop to a high-yielding food staple that is increasingly being converted into *gari* (a dry cereal-like product) and consumed by rural and urban people in West Africa. For the cassava transformation to advance to the next stage—livestock feed and industrial raw material—labor-saving production, harvesting, and processing technologies are needed to reduce costs and improve productivity, and thus reduce the price of cassava products.

# Methods of the COSCA Study

The COSCA study was carried out under the direction of the Project Leader, Professor Felix I. Nweke. The study was administratively based at the Resource and Crop Management Division (RCMD), International Institute of Tropical Agriculture (IITA) Ibadan, Nigeria. The COSCA project was designed to collect farm-level information on cassava over a wide area in Africa in order to generate information to guide research, extension, and policy interventions. The initial budget was US$2.2 million but the amount actually spent was around US$3.0 million. The study was executed over a period of eight years from 1989 to 1997 in six African countries: Congo, Côte d'Ivoire, Ghana, Nigeria, Tanzania, and Uganda. The countries were purposely selected to cover the leading cassava-producing countries in West, Central, and East Africa.

The COSCA study was carried out in phases. The Phase I, village-level survey, compiled information for generating hypotheses and sample frames for subsequent surveys. Phase II, plot-level surveys, were a detailed study of cassava fields. Phase III, post-production surveys, included a household, marketing, and processing studies.

### The Study Team

The COSCA Steering Committee of nine scientists was drawn from African national and international research and development agencies (see appendix 2). The Steering Committee defined guidelines for the study, problems, survey designs, and methodologies for data collection and analysis.

In each country, field data collection was carried out by a team of national scientists who were selected from the senior staff of the cassava research and extension programs of that country. This represents a departure from the common practice of using secondary school graduates to interview farmers and collect farm-level data in Africa. Yet, using senior scientists to interview farmers and collect farm-level data had several advantages: they collected high-quality data because they are experienced researchers; their research work benefitted from their exposure to farmers and farm-level situations; and they were able to apply the knowledge from the COSCA study before the study reports were prepared.

Scientists from the various IITA research divisions contributed to the COSCA study on a part-time basis. The team also included part-time subject matter specialists in marketing, geography, food technology, nutrition, health, and anthropology from the following collaborating international agencies: the Centro Internacional de Agricultura Tropical (CIAT), Cali, Columbia; the International Child Health Unit, Uppsala University, Uppsala, Sweden; the Natural Resources Institute (NRI), London; and the Rockefeller Foundation.

### Sampling Procedure

For each country, the village sample frame was generated using the Geographic Information System (GIS) method (Carter and Jones 1989). The village sample frame consisted of a stratification of the cassava-producing areas of each country by climatic (agro-ecology), population density, and market access infrastructure conditions. The climatic zones were defined

**Table App. 1.1.** Principal Climatic Classes for Cassava in Africa. *Source: Carter and Jones 1989.*

| CLASS | MEAN GROWING SEASON TEMPERATURE | NO. OF DRY MONTHS[a] | GROWING SEASON TEMPERATURE RANGE |
|---|---|---|---|
| Lowland Humid | > 22°C | < 4 | <10°C |
| Lowland Semi-Hot | > 22°C | 4–6 | <10°C |
| Lowland Continental | > 22°C | 4–6 | 10°C+ |
| Lowland Semi-Arid | > 22°C | 6–9 | 10°C+ |
| Highland Humid | < 22°C | < 4 | 10°C+ |
| Highland Continental | < 22°C | 4–6 | 10°C+ |

[a] A dry month has less than 60 mm of rain

as in table App. 1.1 and regrouped, for the purpose of data analyses and discussions, as the forest (i.e., lowland and highland humid classes), transition (i.e., lowland semi-hot, lowland and highland continental classes), and savanna (i.e., lowland semi-arid class) agroecological zones (fig. App. 1.1)

Areas with population density of fifty persons and above per square kilometer were classified as high population density zones and areas with fewer than fifty persons were classified as low-density zones. The population data were from the United States Census Bureau. Market access infrastructure zoning was based on all-weather roads, railways, and navigable rivers from Michelin travel maps of Africa. Areas with all-weather roads, railways, or navigable rivers were classified as good market access zones and areas without were classified as poor market access zones.

The maximum number of strata (zones) per country was twelve consisting of a factorial combination of three agro-ecological zones (the forest, transition, and savanna), two population density (high and low) zones, and two market access (good and poor) zones. All available zones in a country were studied, except those with fewer than 10,000 hectares of cassava. A predetermined number of villages for each country (table app. 1.2) was allocated to the zones in proportion to the sizes of the zones. The number of villages per country was predetermined on the basis of the physical size of the cassava-producing areas of each country. Each of the

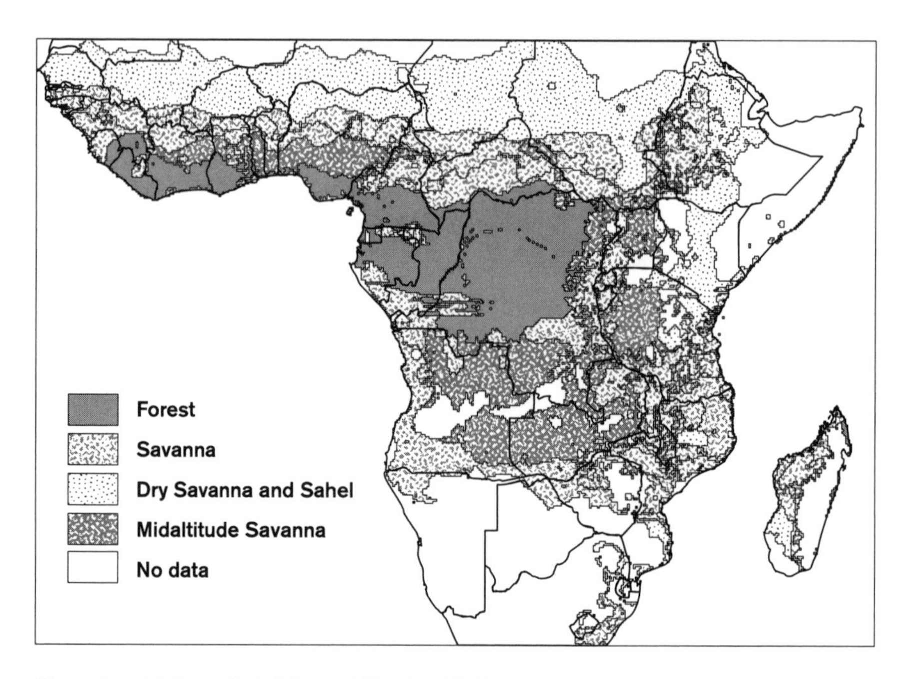


**Figure App. 1.1.** Agroecological Zones of Humid and Subhumid Tropical Africa. *Source: IITA.*

zones was divided into twelve-foot grids. Grids equal in number to the allocated number of villages were selected by a random method in each zone. The grid map was transposed to a detailed regional map using coordinates. One village was selected in each of the selected grids by a random method.

The Phase II (plot-level) production survey sample consisted of three households in each village and all the plots of cassava and other crops cultivated by each of the three households. In each selected village, key informants assisted the COSCA researchers in drawing up a comprehensive list of farm households which cultivated fewer than 10 hectares of all food crops and grouped them into 'large,' 'medium,' and 'small' smallholder units. One household was selected from each group by a random method for the Phase II (plot level) survey. All the fields of a selected household under cultivation with food crops (not only cassava) in the current year were surveyed. The Phase III (post-production) surveys consisted of six households and three market agents per marketplace in

**Table App. 1.2.** Number of Survey Villages, Households, and Fields in the Six COSCA-Study Countries.
*Source: COSCA Study.*

| COUNTRY | NO. OF VILLAGES | NO. OF HOUSEHOLDS | | NO. OF FIELDS PHASE II |
|---|---|---|---|---|
| | | PHASE II | PHASE III | |
| Congo | 71 | 108 | – | 264 |
| Côte d'Ivoire | 40 | 120 | 222 | 267 |
| Ghana | 30 | 90 | 180 | 297 |
| Nigeria | 65 | 195 | 359 | 975 |
| Tanzania | 39 | 131 | 252 | 543 |
| Uganda | 39 | 120 | 240 | 262 |

each village. The six households were the three Phase II households plus three other ones selected by a random method from the village household sample frame.

The plot-level surveys in phase II and post-production surveys in Phase III were separated for two reasons. First, if both phases were combined into one survey it would have been too long and would have tired both the enumerators and respondents. Second, the study was under immense pressure to provide the plot-level information for breeders at IITA to help them define cassava breeding objectives.

Phases II and III surveys were carried out in the Phase I villages. Each of the three phases was implemented simultaneously in each of the six COSCA study countries in the same villages and households between 1989 and 1992. Hence, the surveys in the six COSCA study countries constitute independent studies that were carried out on the same samples and with a common survey design.

A maximum of three marketplaces where farmers in a selected village sold cassava products were selected by a random method. A general market characterization preceded a survey of market agents in each market. The market characterization involved identification of available marketing facilities, organizations, and services and a listing of market agents who traded each cassava product. Three agents for each market were selected by a random method from the list of the agents.

### Survey Instruments

The ease of administration of the survey instruments by the enumerators and the patience of the respondents were of utmost concern in the design of the survey instruments. The questions were written as simply as possible to reduce multiple interpretations. The questions were mostly pre-coded. The list of codes for each question was placed as close to each question as possible.

The Phase I (village-level) survey instrument was divided into three volumes because it was lengthy. The instrument for the household part of the Phase III (post-production) surveys was similarly divided into three volumes. One was for household roster, another was for household processing, and the third was for household expenditure, food consumption, and nutrition. A draft version of each of the survey instruments was extensively pretested before it was finalized and the survey implemented. The Phase I (village-level) survey instrument was pretested in twenty-five of the sixty-five Nigerian survey villages; the Phase II (plot-level) survey instrument was pretested in nineteen of the thirty-seven Ugandan survey villages; and the Phase III (post-production) survey instrument was pretested in fifteen of the sixty-five Nigerian survey villages.

A strict code of conduct was maintained in dealings with the farmers and the traders. Visits were only by appointments and efforts were made to keep appointments on time, although it was not always possible because of unpredictable travel conditions. A local person, an agricultural extension agent or local government official, was engaged to introduce the survey team to the head of a village, who was identified by the survey team independent of the local government official. The local government official was paid at this stage, and the village head introduced the survey team to individual farmers.

The COSCA vehicle carried the logo of a national research or extension agency, and the team members were introduced in the villages as employees of the national agencies even when a visiting international scientist was with the team. The team members dressed modestly. When ever possible, the team members spoke in the local language and dialect.

Local protocols were respected. The team paid a courtesy call to the village head, and if he insisted, he was interviewed even when he was not in the sample. The team patiently extended greetings, prayers, and hospitalities. Gifts which included prepared food, drinks, and, in some cases, livestock, were accepted. If the food or the drink did not appear wholesome, each team member still tested it and expressed appreciation with explanation that the team members were not hungry or thirsty. When the gift was too elaborate, the team accepted the gift with protest.

The only gifts the team was permitted to present to the respondents were planting materials or literature which had been requested by the farmers and promised by the team in a previous visit. However, the farmers' gifts to the teams were sometimes expensive. For example, in many villages, the farmers provided free board and lodging for all team members for the entire period the team was in the village. The team was encouraged to return such a favor at the end of the visit with cash from their per diem with clear explanation that the cash was not a payment for the information the farmers had already provided but a reciprocation of the farmers' hospitality.

For the Phase I (village-level) survey, three different farmer groups were interviewed simultaneously in a village with the three volumes of the survey instrument to reduce the length of time an individual farmer being interviewed was engaged. Each farmer group consisted of a minimum of five men and five women of different ages. Visual aids such as plant specimens and drawings were used to conceptualize issues and facilitate the farmers' understanding of the questions. The farmers were encouraged to discuss the question among themselves before responding to a question. These measures helped to sustain group interest.

For the Phase II (plot-level) survey, the investigating team camped in a village for three days and completed the surveys of a farmer's fields before moving on to another farmer. The Phase II (plot-level) survey was labor-intensive for both the team of enumerators and the farmers and their spouses, as the survey entailed plot perimeter measurements, cassava yield measurement, and soil sampling in addition to the farmers' interviews. In most cases, the farmer and his or her spouse were engaged all day for a number of days in conducting the enumerators to and around each of the farmer's several plots, in taking yield measurements and soil samples, and in responding to the interview questions. Because of this situation, the farmers and their spouses were paid the going farm wage for the village for the number of days the farmers were so employed. Where plots were owned individually by household members, the individual plot owner responded to the interview questions. The interviews were conducted in the field.

The different volumes of the Phase III (post-production) survey instrument were administered to a household in series because the head of the household and his spouse were required to provide answers to questions together. A survey team of three camped in a village for six days; one team member was responsible for administering one volume of the instrument to all of the six households in the village. In this way, the survey team members rotated among the households with different sets of questions while the households had gaps in time between different interviews to attend to their other matters.

In the case of the market survey, the potential was high for distraction because sellers were advertising their wares to potential buyers. Buyers were constantly haggling with the sellers for lower prices. Everyone was constantly exchanging greetings with acquaintances. The team of investigators was very patient with the respondents, allowing them to engage in the advertising, haggling, and exchange of greetings while being interviewed.

At the end of each interview, the respondents were given ample opportunity to ask the researchers questions. Honest answers were provided to the questions by the team of investigators. The researchers avoided making promises that could not be kept. The most common questions of the respondents were what the advantages of the study were to them individually or to their respective villages and whether the visiting team could help attract government supplies of seeds, fertilizers, credit, or access roads. The researchers responded in all cases that the benefit of the study would accrue to all farmers of Africa and not to individual sample farmers or villages, and that if and when a benefit of the study reached the village it might not be easily associated with the visit. In this respect, the time the farmers spent with the researchers and the information the farmers provided were sacrifices on the farmers' part for the benefit of all farmers of Africa. The farmers were told in very clear terms that the COSCA researchers could not influence government provision of seeds, fertilizers, credit, or access roads. The researchers, however, provided, where applicable, planting materials of improved varieties of not only cassava but also of other crops, if available, in subsequent visits.

### Logistics and Supervision

Each of the surveys was preceded by a two-week planning meeting of all the collaborating team members at the IITA in Ibadan. For the Phase II (plot-level) and Phase III (post-production) surveys, the first week of the planning meeting was devoted to data cleaning and to preliminary analyses of data and experiences gained from the previous survey that might influence the next survey. The second week was devoted to the planning of the next survey. The planning focused on identification of the problems, data needs, and sources, and the best way to obtain the data needed, logistics of travel and supervision. Plans for supervision were prepared and presented with emphasis. A four-wheel-drive vehicle was provided for each country team.

The Phase I (village-level) and Phase II (plot-level) surveys were intensively and painstakingly supervised in the six COSCA study countries by the Project Leader. Intensive supervision helped reduce undue delays in the survey. In some of the countries, the national team members were often apathetic or had other priorities. In such countries, the national teams were often coerced into going out to the field for the survey only by a planned supervision visit. The tendency for the national team members to be dishonest with respect to work or money was reduced by their awareness that the Project Leader, having been to the survey fields, could not be easily deceived.

Despite these precautionary measures, however, there were cases of bad data which were due to errors or mistakes. Whenever the bad data problem was suspected, the survey was repeated by a different set of researchers. The Phase I (village-level) survey was repeated in some villages in each of the six COSCA study countries. The Phase II (plot-level) survey was repeated in some Nigerian and Tanzanian villages and was discarded in its entirety for all Ugandan villages. The plot-level information analyzed for Uganda was that collected in the elaborate pretest of that survey phase in that country. The market survey was repeated for all Ghanaian villages. The rest of the Phase III survey information from most of the villages in Ghana and all the Phase III survey information from several of the villages in the Côte d'Ivoire are poor but the surveys could not be repeated because of a lack of funds.

National salaries were very low and irregular at the time of the COSCA study because the economies of most African countries had started to deteriorate. In most countries, public servants moonlighted to earn additional incomes. To discourage the national COSCA team members from engaging in other moneymaking activities while the COSCA study was in progress, they were given a monthly stipend of US$200, which was often several times larger than their monthly government wages, and the stipend was paid on schedule. In addition, per diem was fixed at the national government rate, which was, in all cases, significantly higher than the United Nations' rate that the IITA uses to pay to its international scientists for the same rural locations in Africa. The national governments rarely paid the high per diem because the staff traveled infrequently, and even when they traveled they were not sure to be paid.

## Data Processing

The national team members conducted the field surveys and shipped the completed questionnaires to Nigeria for transcription. The plan was for the national team members to analyze the data and prepare the country reports, but because of numerous problems, the country reports were prepared in Ibadan.

The Project Leader knew from an earlier study that the COSCA study required an elaborate arrangement to effectively cope with the handling of the large amount of data which the study entailed. Therefore, a data processing-group was set up at the IITA to transcribe and analyze the data in Ibadan. The group included two programmers and four data analysts, all of whom were university graduates of computer science, and four data entry clerks. The unit was headed by Mr. Sunday Folayan, an electrical engineering M.S. degree holder.

The data-processing personnel participated in the various COSCA planning meetings and in the design of the survey instruments and in precoding of the questionnaires. The programmers transformed the precoded questionnaires into a computer format to enable the data to be transcribed directly into the questionnaire format in the computer.

The COSCA researchers were conscious of data quality control at every stage of the study, but quality of data control was a critical issue at the data transcription stage. The four data clerks transcribed all the data for all the surveys. The data transcription was labor intensive but the proficiency, in terms of speed and accuracy, of the clerks improved with time. The four clerks first transcribed data independently, taking frequent breaks to reduce fatigue. Then they paired up to check for errors. One person would read from the questionnaire and the other from the computer, and the pair agreed that the two tallied for every figure. They were not left alone, and the Project Leader was on hand to clarify any doubts for the clerks. Issues that the Project Leader could not clarify were flagged and referred to the national team members who had collected the data. The data transcription for each survey was completed before the a planning meeting for the next survey which the national team members attended.

Data cleaning was not completed until reports were prepared from the data set. When doubtful analytical results were identified at the report preparation stage (appendix 3), they were checked by going back to the questionnaire and in some cases going back to the farmers.

The analytical team set up in 1990 stayed together until 1996, and its proficiency improved significantly with time. In 1996, the head of the analytical team, Mr. Sunday Folayan, decided to leave for more challenging work. By then, the COSCA data handling had been reduced to routine analyses, and members of the analytical team had gained tremendous experience in the previous six years (1990 to 1996). Mr. Folayan requested the Project Leader for US$1,500 (as an investment) to set up an Internet services provider business in Ibadan. This partnership gave birth to the General Data Rngineering Services (Nigeria) Limited (GDES), which has its headquarters in Bodija, Ibadan. Professor Nweke is chairman of board of directors and Mr. Folayan is the executive director of operations.

# List of COSCA Collaborators

### Executive Agency

International Institute of Tropical Agriculture (IITA), Ibadan, Nigeria

### Collaborating Agencies

Centro Internacional de Agricultura Tropical (CIAT), Cali, Colombia
Natural Resources Institute (NRI), London
International Child Health Unit (ICHU), Uppsala University, Uppsala
National Agricultural Research Systems (NARs) of the Congo, the Côte d'Ivoire, Ghana, Nigeria, Tanzania, and Uganda

### Primary Funding Agency

The Rockefeller Foundation (RF)

### Steering Committee

D. Spencer, IITA (Chairman)
R. Barker, IITA Board of Trustees
R. Best, CIAT
M. Dahniya, University of Sierra Leone
S. Hahn, IITA

J. Lynam, RF
T. Phillips, University of Guelph
F. Quin, IITA
J. Strauss, Michigan State University

**Project Leader**

F. Nweke, IITA

**National Coordinators**

N. Lutaladio, Congo
V. Kuezi-Nke, Congo
B. Bukaka, Congo
N. Kalombo, Congo
D. Lutete, Congo
K. Tano, Côte d'Ivoire
K. Ngora, Côte d'Ivoire
M. Ehui (Ms), Côte d'Ivoire
R. Al-Hassan (Ms), Ghana
J. Otoo, Ghana
S. Asuming-Brempong, Ghana
A. Kissiedu, Ghana
A. Cudjoe, Ghana
J. Haleegoah (Ms.), Ghana
B. Ugwu, Nigeria
O. Ajobo, Nigeria
C. Asinobi (Ms.), Nigeria
J. Njoku, Nigeria
G. Okwor, Nigeria
R. Kapinga (Ms.), Tanzania
B. Rwenyagira, Tanzania
E. Ruzika, Tanzania
G. Otim-Nape, Uganda
A. Bua, Uganda
Y. Baguma, Uganda

### IITA Scientists

R. Asiedu, Breeder (TRIP)

M. Bokanga, Biochemist (TRIP)

A. Dixon, Breeder (TRIP)

### Non-IITA Scientists

S. Carter, Geographer (CIAT)

N. Poulter, Food Technologist (NRI)

A. Westby, Biochemist (NRI)

J. Lynam, Agricultural Economist (RF)

S. Romanoff, Anthropologist (RF)

H. Rosling, Physician (ICHU)

### Consultants

M. Akoroda, University of Ibadan

C. Asadu, University of Nigeria, Nsukka

L. Fresco (Ms), Wageningen University

W. Jones, Stanford University

E. Okorji, Univ. of Nigeria, Nsukka

### Data Processing Group

S. Folayan

O. Lawal

E. Oyetunji

A. Adeyeye

A. Akintunde

O. Brodie-Menz

### Notes

PHMD = Plant Health Management Division

RCMD = Resource and Crop Management Division

TRIP = Root and Tuber Improvement Program

# List of COSCA Reports

**Working Papers**

No. 1. Nweke, Felix I. 1988. COSCA project description. IITA (International Institute of Tropical Agriculture, Ibadan, Nigeria).

No. 2. Carter, S. E., and P. G. Jones. 1989. COSCA site selection procedure. IITA.

No. 3. Nweke, Felix I., John Lynam and Coffi Prudencio, eds. 1989. Status of data on cassava in major producing countries of Africa: Cameroon, Côte d'Ivoire, Ghana, Nigeria, Tanzania, Uganda, and Zaire. IITA.

No. 4. Nweke, Felix I., John Lynam and Coffi Prudencio, eds. 1990. Methodologies and data requirements for cassava systems study in Africa. IITA.

No. 5. Stoorvogel, J. J., and L. O. Fresco. 1991. The identification of agro-ecological zones for cassava in Africa, with particular emphasis on soils. IITA.

No. 6. Ugwu, B. O., and P. Ay. 1992. Seasonality of cassava processing in Africa and tests of hypotheses. IITA.

No. 7. Natural Resources Institute (NRI). 1992. COSCA Phase I processing component. IITA.

No. 8. Berry, Sara S. 1993. Socio-economic aspects of cassava cultivation and use in Africa: Implications for the development of appropriate technology. IITA.

No. 9. Fresco, Louise O. 1993. The dynamics of cassava in Africa: An outline of research issues. IITA.

No. 10. Nweke, Felix I., A. G. O. Dixon, R. Asiedu, and S. A. Folayan. 1994. Cassava varietal needs of farmers and the potential for production growth in Africa. IITA.

No. 11. Nweke, Felix I. 1994. Processing potential for cassava production growth in sub-Saharan Africa. IITA.

No. 12. Nweke, Felix I. 1994. Cassava distribution in sub-Saharan Africa. IITA.

No. 13. Nweke, Felix I. 1996. Cassava production prospects in Africa. IITA.

No. 14. Nweke, Felix I. 1996. Cassava: A cash crop in Africa. IITA.

No. 15. Nweke, Felix I., B. O. Ugwu, and A. G. O. Dixon. 1996. Spread and performance of improved cassava varieties in Nigeria. IITA.

No. 16. Nweke, Felix I., R. E. Kapinga, A. G. O. Dixon, B. O. Ugwu, and O. Ajobo. 1998. Production prospects for cassava in Tanzania. IITA.

No. 17. Nweke, Felix I., G. W. Otim-Nape, A. G. O. Dixon, O. Ajobo, B. O. Ugwu, A. Bua, Y. Baguma, and H. Masembe-Kajubi. 1998. Production prospects for cassava in Uganda. IITA.

No. 18. Asadu, C. L. A. and Felix I. Nweke. 1998. The soils of cassava-growing areas in sub-Saharan Africa. IITA.

No. 19. Nweke, Felix I., and Anselm A. Enete. 1998. Gender surprises in food production, processing, and marketing by gender with emphasis on cassava in Africa. IITA.

No. 20. Nweke, Felix I., B. O. Ugwu, A. G. O. Dixon, C. L. A. Asadu, and O. Ajobo. 1998. Cassava production in Nigeria: A function of farmer access to markets and to improved production and processing technologies. IITA.

No. 21. Nweke, Felix I., J. Haleegoah, O. Ajobo, B. O. Ugwu, A. G. O. Dixon, and R. Al-Hassan. 1998. Cassava production in Ghana: A function of market demand and farmer access to improved production and processing technologies. IITA.

No. 22. Nweke, Felix I., D. Lutete, A. G. O. Dixon, B. O. Ugwu, O. Ajobo, N. Kalombo, and B. Bukaka. 1998. Cassava production and processing in the Democratic Republic of Congo. IITA.

No. 23. Nweke, Felix I., K. N'Goran, A. G. O. Dixon, B. O. Ugwu, O. Ajobo, and T. Kouadio. 1998. Cassava production and processing in Côte d'Ivoire. IITA.

**Ph.D. Dissertations**

1. Ezedinma, Chukwuma Ike. 1993. Farm labor availability and use in food crop production in the cassava producing zones of tropical Africa. University of Nigeria.
2. Ndibaza, Regina Ernesti. 1994. Intercropping of cassava (*Manihot esculenta Crantz*) and sweet potato (*Ipomea batatas L.*) in the semi-arid zone of Tanzania. University of Ibadan.
3. Nwajiuba, Chinedum Uzoma. 1995. Socioeconomic impact of cassava post-harvest technologies in southeast Nigeria. University of Hohenheim.
4. Tshiunza, Muamba. 1996. Agricultural intensification and labor needs in the cassava-growing zones of sub-Saharan Africa. Katholieke Universiteit Leuven.
5. Asinobi, Chinagorom Onyemaechi. 1998. Nutritional status of preschool children in cassava producing areas of Nigeria as assessed by nutritional anthropometry and food intake. University of Ibadan.
6. Lemchi, Jonas Ibe. 1999. The marketing system for cassava in Nigeria. Federal University of Technology, Owerri, Nigeria.
7. Camara, Yousouf. 2000. Profitability of cassava production systems in West Africa: A comparative analysis (Côte d'Ivoire, Ghana, and Nigeria). Michigan State University.
8. Johnson, Michael Emmet. 2000. Adoption and spillover of new cas-

sava technologies in West Africa: Econometric models and heterogenous agent programming. Purdue University.

9. Enete, Anselm. Cassava marketing in sub-Saharan Africa. Katholieke Universiteit Leuven (in preparation).

# Demand Function Specifications

The demand equations specified are based on the Working (1943) and Lesser (1963) model:

$$\omega_i = \alpha_i + \beta_{1i} \log M + \gamma_i \log n + \sum_d \gamma_{id} \left(\frac{n_d}{n}\right) + \varepsilon_i \quad (1)$$

where, subscript $i$ refers to the commodity and $M$ refers to annual household per capita expenditure. Defining $m_i$ as total household expenditure on commodity $i$ and $E$ as total household expenditure yields $\omega_i = m_i/E$ as the budget share of commodity $i$, out of total household expenditure.

This flexible functional form allows for non-linearities in income and household composition and is readily derived from demand theory. Demographic effects in consumption are incorporated by including household size in logarithms, the ratio of household members in a particular demographic group ($n_d$), out of total household size ($n$)—for $d$ = 1 to 10 demographic groups. Table 2 describes these demographic groups. Village dummies were used to capture the effects of prices and other village level effects on demand. Parameters ($\beta_{1i}$, $\gamma_i$, $\gamma_{id}$) estimated from equation (1) are then used to estimate the expenditure and outlay equivalent elasticities. The formula for the expenditure elasticity of demand is:

$$\eta_l = \frac{\beta}{\omega} + 1 \tag{2}$$

where subscript $l$ refers to the linear specification of per capita expenditure.

Even though the above specification is flexible, it still restricts the sign of the marginal propensity to consume such that it is increasing for luxuries ($\eta_l > l$), and decreasing for goods with $\eta_l < 1$ (Thomas, Strauss, and Barbosa 1991). In order to relax this restriction, we also specify a quadratic demand function:

$$\omega_i = \alpha_i + \beta_{1i} \log M + \beta_{2i} (\log M)^2 + \gamma_i \log n + \sum_d \gamma_{id} \left(\frac{n_d}{n}\right) + \varepsilon_i \tag{3}$$

The resulting expenditure elasticity is:

$$\eta_q = \left(\frac{\beta_1 + 2\beta_2 \log E}{\omega}\right) + 1 \tag{4}$$

The above specification also allows one to estimate the effect of household composition on per capita expenditure. Generally, since the items examined here are food staples, one would expect that an increase of one more household member of category $j$ would lead to an increase in food staple expenditure. Deaton et. al. (1989) and Deaton (1997) derive an elasticity measure which allows one to examine this effect. They call this measure *outlay equivalents*. It is the fraction by which per capita expenditure would have to increase (decrease), in order to induce the same increase (decrease) in expenditure on good $i$ as would an additional household member of category $j$. The equation is as follows for the linear specification, respectively:

$$\pi_{ij} = \frac{\partial q_i / \partial n_j}{\partial q_i / \partial E} \div \frac{E}{n} = \frac{(\gamma_i - \beta_{1i}) + \gamma_{ij} - \sum_d \gamma_{id}(n_d/n)}{\beta + \omega} \tag{5}$$

where $\pi_{ij}$ is the outlay-equivalent ratio for good $i$ and demographic category $j$.

### Variable Description and Construction

Table 4 summarizes the variables used in estimating equations (1) and (3). Expenditure shares were estimated as a share of total household expenditure, and household expenditure was estimated as the sum of food, nonfood, and rent. Estimation of food and nonfood expenditures were straightforward. However, in the case of the rent variable, preliminary examination revealed that only a small percentage of households were actually renting. As a result, this variable could not be used in estimating rental expenditure. Instead, an alternative variable was estimated. This variable was based on the household's estimate of the resale value of their dwelling. Thus, using data on resale value of the dwelling and also data on the number of years till the next major repairs on the dwelling, an annual depreciation rate was estimated. This variable was then regressed on: number of rooms in the dwelling, material of outer wall, access to portable water, electricity, toilet facilities, and village dummies. Various functional forms were specified. The functional form which provided the best result for each country was then used to predict an estimate of rent or mortgage payment for each household.

**Table App. 4.1.** Means and Standard Deviations of Variables Used in Regression Analysis.

| | COTE D'IVOIRE | | GHANA | | NIGERIA | | TANZANIA | | UGANDA | |
|---|---|---|---|---|---|---|---|---|---|---|
| | MEAN | σ | MEAN | σ | MEAN | σ | MEAN | σ | MEAN | σ |
| **Food Share** | .706 | .187 | .737 | .171 | .721 | .160 | .781 | .168 | .761 | .165 |
| **Cassava** | | | | | | | | | | |
| All | .119 | .144 | .128 | .130 | .109 | .099 | .091 | .131 | .178 | .171 |
| Fresh Root | .082 | .114 | .091 | .111 | .016 | .036 | .034 | .069 | .134 | .146 |
| Granules | .017 | .049 | .008 | .012 | .041 | .067 | .0001 | .003 | 0 | 0 |
| Paste | .009 | .041 | .017 | .058 | .025 | .055 | 0 | 0 | 0 | 0 |
| Chips | .009 | .039 | .011 | .025 | .026 | .053 | .056 | .114 | .043 | .095 |
| **Yam** | .059 | .118 | .132 | .179 | .109 | .124 | .003 | .030 | .007 | .016 |
| **Sweet Potato** | 0 | 0 | .001 | .005 | .004 | .015 | .023 | .060 | .041 | .067 |
| **Banana/Plantain** | .053 | .070 | .030 | .066 | .009 | .037 | .033 | .080 | .054 | .079 |
| **Pulses** | .0002 | .002 | .009 | .018 | .034 | .037 | .057 | .076 | .068 | .075 |
| **Wheat** | .00198 | .0123 | .0018 | .025 | .00065 | .0041 | .0055 | .024 | .0014 | .0078 |
| **Grains** | | | | | | | | | | |
| All | .032 | .079 | .056 | .089 | .104 | .111 | .124 | .153 | .067 | .089 |
| Maize | .0316 | .078 | .052 | .081 | .059 | .062 | .102 | .135 | .039 | .065 |
| Sorghum/Millet | .0006 | .008 | .004 | .030 | .044 | .077 | .022 | .071 | .028 | .065 |
| **Rice** | | | | | | | | | | |
| All | .085 | .1078 | .016 | .035 | .047 | .051 | .074 | .117 | .009 | .028 |
| Local | .059 | .098 | .010 | .028 | .042 | .051 | .065 | .113 | .009 | .028 |
| Imported | .025 | .069 | .006 | .024 | .004 | .018 | .009 | .041 | .0005 | .004 |
| **Log Per Capita Expenditure** | 7.40 | .851 | 8.13 | .741 | 4.86 | .714 | 6.7 | .86 | 8.37 | .715 |
| **Per Capita Expenditure Log Household Size** | 2.056 | .645 | 1.98 | .605 | 2.23 | .645 | 1.89 | .609 | 1.86 | .563 |
| **Household Size** | 9.49 | 6.24 | 8.55 | 5.06 | 11.29 | 7.25 | 7.80 | 4.59 | 7.47 | 4.18 |
| **Males 0–5** | .0914 | .109 | .0916 | .113 | .0739 | .093 | .0876 | .115 | .117 | .130 |
| **Females 0–5** | .116 | .121 | .0897 | .104 | .0654 | .0827 | .0931 | .119 | .093 | .117 |
| **Males 5–14** | .109 | .127 | .136 | .140 | .115 | .117 | .115 | .119 | .123 | .145 |
| **Females 5–14** | .103 | .115 | .108 | .115 | .117 | .109 | .110 | .124 | .127 | .132 |
| **Males 14–55** | .192 | .149 | .259 | .184 | .219 | .149 | .199 | .164 | .228 | .160 |
| **Females 14–55** | .240 | .144 | .231 | .130 | .263 | .142 | .247 | .149 | .232 | .133 |
| **Males 55–65** | .0327 | .067 | .021 | .091 | .031 | .075 | .0241 | .0609 | .027 | .091 |
| **Females 55–65** | .038 | .105 | .0199 | .058 | .0249 | .0702 | .0347 | .108 | .011 | .050 |
| **Males 65** | .0358 | .104 | .0211 | .0633 | .048 | .110 | .0511 | .127 | .0221 | .088 |
| **Females 65** | .0656 | .144 | .0319 | .0735 | .0743 | .153 | .077 | .166 | .0394 | .518 |
| **Observations** | 220 | | 180 | | 360 | | 285 | | 240 | |

# Notes

### Chapter 1

1. Based on average per capita calorie consumption per year.
2. *Gari* is a granulated and toasted cereal-like cassava food product that is convenient for consumption in urban environments because it is in a ready-to-eat form and has an extended shelf life.
3. Cassava is called manioc in Portuguese.
4. This section draws heavily on Jones (1959) and Okigbo (1980).
5. The period before seasonal food crops are ready for harvest.
6. Cassava does not have a period of maturity. As the plant grows the root continues to bulk (swell), until after a period of three or four years, when deterioration begins.

### Chapter 2

1. The physical soil properties assessed are clay, silt, and sand content. The chemical properties assessed are total nitrogen (Total N), organic matter (OM), available phosphorous (Available P), calcium (Ca), magnesium (Mg), potassium (K), sodium (Na), and manganese (Mn) contents; total exchangeable bases (TEB), total exchangeable acidity (TEA), effective cation

exchange capacity (ECEC), base saturation, and soil pH. For more details see Asadu and Nweke 1999.

2. Several of the fields have been under continuous cultivation of cassava for a minimum of a decade.
3. The agro-ecological zones of these countries include the humid forest (hereafter forest) zone, the forest-savanna transition (hereafter transition) zone, and the moist savanna (hereafter savanna) zone. The forest zone is characterized by nine to twelve months of growing season, more than 22°C daily mean temperature, and less than 10°C temperature range. The transition zone is characterized by five to seven months of growing season, above 22°C daily mean temperature, and above 10°C temperature range. The savanna zone is characterized by three to four months of growing season, above 22°C daily mean temperature, and above 10°C temperature range (Carter and Jones 1989).
4. Unpublished data from S. K. Hahn, IITA, Ibadan.
5. Hahn is Dr. S. K. Hahn unless otherwise specified.
6. Cassava is normally grown on flat land in sandy soils and on ridges and mounds in heavy lateritic soils (S. K. Hahn 1984).
7. Ownership of the main crop in the field was determined from responses to the question "Who in the household owns this field?" The question was addressed to the head of the household and the spouse, where applicable. Possible responses were husband, son, other male member, wife, daughter, other female member, and all members jointly.
8. The soil fertility status of the women's and the men's cassava fields was not assessed in the COSCA study.
9. The customary "sweet" and "bitter" cassava varieties depend upon the amount of cyanogens (prussic acid) in the edible parts of the roots (Jones 1959, 12). The roots of sweet cassava are low in cyanogens, mealy after cooking, and usually eaten raw, boiled, or roasted in an open fire. Bitter cassava varieties are high in cyanogens, waxy after cooking, and are harmful to humans and animals unless they are peeled, grated, and toasted or soaked in water for a few days and boiled or sun-dried.
10. A human disorder characterized by a gradual onset of sensory deficit in the feet, gait abnormality, and visual impairment in middle-aged people

(Onabolu 2001).

11. Cassava is prepared either by peeling, grating, fermenting, and toasting or by soaking in water for four or five days and then sun-drying to eliminate cyanogens.

## Chapter 3

1. Johnston (1958, 226) described rice as "the glamour crop" of West Africa.
2. The term "lifeline" is borrowed from the title *Cassava: Lifeline for the Rural Household,* edited by Anthony E. Ikpi and Natalie D. Hahn (1989).
3. Memorandum, Raw Materials Research and Development Council. Abuja, Nigeria.
4. The by-products of cassava processing that are fed to livestock in Africa are most likely to be included as waste in official statistics.

## Chapter 4

1. The shape of cassava leaves is either broad or narrow. Pubescence refers to the presence of hairiness in the growing tip of the plant; its presence is determined with the aid of a magnifying lens. All color characteristics are assessed with the aid of a color chart: the young shoot is either green or purple; the petiole is red or green; the inner skin is white, cream, or red; and the root flesh is white, cream, or yellow. The petiole connects the leaf to the stem; it is considered red if it has any red spot. Otherwise, it is considered green.
2. The procedure used was the method of principal component analysis (Pearson 1901; Hotelling 1933). The 1,200 local varieties are not unique cassava varieties because they are distinguished by locally given names. A cluster does not represent a unique cassava variety because there are some important morphological characteristics that are not included in the analysis.
3. Taste is a subjective concept; it will be unsound to base the system of cassava classification on it. Besides, no definite line can be drawn between bitter and sweet cassava varieties because several intermediate-tasting varieties exist, one variety grading into the other. The environment also has an effect on the level of cyanogens in cassava; drought and low soil fertility are known to increase the cyanogen level in cassava. Varietal differences

in vegetative characteristics are more useful in identifying cassava varieties (Doku 1966).

4. The yield losses presented here for the pests and diseases are guesstimates. The interactions among the various pests and diseases, the influences of soil fertility, seasonal factors, the cassava varietal factors, and cropping practices complicate the assessment of yield loss due to specific pests and diseases (Thresh et al. 1997).
5. In the early 1990s, an epidemic of an extremely severe form of the mosaic disease spread through most of Uganda. Researchers discovered that the virus epidemic was caused by a new form of cassava *gemini* virus. The Uganda variant of the *gemini* virus is now widely distributed in Uganda (Harrison et al. 1997).
6. In the COSCA study, the assessment of the importance of cassava relative to other staples is in terms of area planted to cassava as a percentage of the area planted to all staples (Jones 1959).
7. The Lake Tanganyika zone is a cassava-growing area of Tanzania not covered in the COSCA study.
8. This figure is for root yield as distinct from leaf yield; in the Congo and Tanzania cassava leaves are harvested and eaten as a vegetable.
9. There is a positive relationship between stand density and yield. Farmers in high population density villages plant at high stand densities because of shortage of land.

## Chapter 5

1. This section has benefited significantly from Beck 1980; Hahn, Howland, and Terry 1980; and Jennings 1976.
2. Fresco (1986) reports that the yields at the Kiyaka station in the Congo were high because the experimental plots were on freshly cleared forest land.
3. Tree cassava is believed to be a natural hybrid of Ceara rubber and cassava (Jennings 1976).
4. Cours et al. (1997) reported that a parallel research activity in the 1930s following the same approach was carried out independently by the French at Alatroa agricultural research station in Madagascar and achieved results similar to those at the Amani research station.

5. This is comparable to the six years (1960–66) that it took IRRI researchers to develop the famous IR-8 green revolution rice variety (Ruttan 2001, 77).
6. Hahn (2000) reported that Ms. A. K. Howland was especially helpful in providing information on Storey's research program on the mosaic disease. Ms. Howland was a former colleague of H. H. Storey at the Amani research station in Tanzania. She was invided by S. K. Hahn to join his research team at the IITA from 1972 to 1976.
7. At about the time of the commencement of the IITA's cassava breeding program in 1971, a new and serious disease of cassava, the bacterial blight, was reported in Nigeria. The disease spread to the Congo, Cameroon, Togo, Benin, Ghana, Uganda, Kenya, Burundi, Rwanda, and the Central African Republic.
8. The Ceara rubber x cassava hybrids were not real cassava because they did not stand erect and they produced low root yields that were of poor food quality. S. K. Hahn crossed the Ceara rubber x cassava hybrid (58308) with West African and South American cassava varieties that were susceptible to mosaic but stood erect and gave high root yields that were of good food quality. The result was the mosaic-resistant and high-yielding TMS varieties.
9. The diffusion of the TMS varieties is discussed in detail in chapter 7.
10. The farm-level yield of the high-yielding TMS varieties was not significantly different from the yield in researcher-managed on-farm trials conducted by S. K. Hahn. For example, the yields of the TMS varieties in researcher-managed on-farm trials were 21.0 tons per hectare in 1983, 23.5 tons per hectare in 1984, and 16.0 tons per hectare in 1985 in different locations in the forest zone of Nigeria (International Institute of Tropical Agriculture 1986).
11. We shall show in chapter 6 that farmers wishing to take advantage of the high yield of the TMS varieties adjust the stand density and planting date of the branching cassava in intercropping.
12. The estimation of the cyanogen content in the cassava roots was based on the picric acid test method (Almazan 1987). This method is semiquantitative; it was adopted because an alternative quantitative method suitable for quick assessment in the field was not available.
13. For example, in the Congo, the Programme National du Manioc (PRONAM) was established in 1974; in Nigeria, the National Root Crops Research Insti-

tute (NRCRI) was established in 1975; in the Côte d'Ivoire, the cassava program of the Institut des Savannes (IDESSA) was established in 1979; and in Uganda, the National Root Crop Research Program was established in 1981. These programs were established with IITA's technical assistance and with financial assistance secured with IITA's support from international donor agencies such as United States Aid for International Development (USAID), IFAD, the Gatsby Foundation, and so on.

14. IITA clones were earlier versions of the TMS varieties.
15. In 1989, the monthly salary of a professor at Makerere University in Uganda was equivalent to the cost of seven liters of beer. Even then, his university salary was paid three months late. In 1990, the monthly government salary of the head of the COSCA research team in the Congo was less than his monthly electricity bill. At the same time, his government salary was five months overdue.
16. Further breeding based on Jennings's 5318/34 was reported in Ghana (Doku 1969). The most important work, however, occurred at Moor Plantation, Ibadan, Nigeria (Jennings 1976).

## Chapter 6

1. Yam is grown after fallow because it needs high soil fertility.
2. Farmers harvest a cassava plant, break off the stem to a planting size, and stick it right back into the soil.
3. Cassava planting on the flat seed-beds in well-drained soils is a traditional practice and not an adoption of the IITA's minimum tillage technology.
4. Land clearing and tillage were found to be mechanized in only a small proportion of cassava fields in the five COSCA study countries. Mechanized land clearing for cassava production was common only in Uganda, where it was used in 60 percent of the cassava fields. Mechanization of smallholder agriculture was a main objective of the government policy of trying to create a "modern agriculture" in Uganda (Pingali, Bigot, and Binswanger 1987). Most field tasks for cassava production are done manually, using simple implements such as a hand hoe and a cutlass.
5. Mechanized tillage equipment includes oxen and tractors for plowing, harrowing, and ridging. Mechanized tillage was considered adopted in a field

if the tillage was carried out with animal or machine power either in part or in full. Mechanized transportation was by motor vehicles such as four-wheeled motor vehicles, motorcycles, motorboats, and tractors; by animal-pulled carts; by bicycles; or by human-pushed carts.

Bicycles, human-pushed carts, animal-pulled carts, and motorcycles were often owned by smallholders. Tractors were often available from public (and sometimes private) tractor hire services. Four-wheeled motor vehicles such as buses and taxis were available for hire commercially. Farmers with a large supply of produce for market could generally rent taxis or tractors for transporting produce to market (Pingali, Bigot, and Binswanger 1987).

## Chapter 7

1. During the foreign exchange bidding in September 1986, the value of the Naira dropped from US$1.12 to about US$0.30.
2. The Naira has continued to slide. In February 2001, the value was 124 Naira per US$1.00.
3. Personal interview with Mr. O. A. Edache, Abuja, 15 January 2001.
   Soon after the interview the price of *gari* rose sharply due to the demand for Nigeria's dried cassava roots as livestock feed in Europe, a result of the outbreak of mad cow disease.
4. Bendel State was later divided into Delta and Edo States.
5. The National Accelerated Food Production Program (NAFPP) was established in 1972 to design, test, and extend technological packages for five crops: rice, maize, sorghum, millet, and wheat. Cassava was added in 1974. The program was a cooperative effort among research institutes and state and federal government extension services (Anojulu, Adenola, and Ogunbiade 1982).
6. Hahn was honored with the chieftaincy title of *Ba-ale Agbe* (King of Farmers) by the members of a town in Western Nigeria in recognition of his work in developing the TMS varieties.
7. African farmers usually protect promising cassava plants that strangers have planted in their fields. If they produce a good yield, farmers will take cuttings and plant them.
8. Nigeria was divided into nineteen states at the time.

9. The World Bank admits that the Agricultural Development Project (ADP) was a big failure in Nigeria despite the seemingly successful activities described here (World Bank 1993).
10. *Gari* preparation is divided into three main steps: (1) peeling and washing; (2) grating, pressing, and sieving; and (3) toasting. Only the grating, pressing, and sieving step is mechanized in Nigeria.
11. Farmers were processing local cassava varieties using mechanized graters before the establishment of the IITA's cassava program in 1971.
12. During the early 1960s, some Indian agricultural scientists objected to Norman Borlaug's recommendation that the government of India import high-yielding wheat varieties from Mexico. The Indian minister of agriculture agreed with Borlaug and the government imported seventeen thousand tons of wheat seed from Mexico. This seed formed the foundation of India's Green Revolution.
13. Personal telephone discussion with Dr. S. K. Hahn, 20 March 2001.
14. The literal translation of "Monkey de work Baboon de chop" means "monkey works for baboon to eat" This is a West African pigeon English saying.
15. Dr. O. O. Okoli helped introduce the IITA's high-yielding TMS varieties in Ghana under the IITA's technical assistance program to the cassava project of Ghana's Small-Holder Rehabilitation and Development Program (SRDP).
16. Fitaa means white or pale arm; Gblemo Duade means cassava for milling or processing; and Afisiafi means cassava for everywhere, ubiquitous cassava, or widely adapted cassava (Dixon 2000).

## Chapter 8

1. It is difficult to separate cassava processing from cassava food preparation because some combinations of the cassava processing and food preparation activities lead to final cassava food products which are in ready-to-eat forms. Other combinations of the cassava processing and food preparation activities lead to intermediate products which are stored until the need arises for conversion into ready-to-eat forms.
2. The customary "sweet" and "bitter" cassava varieties depend upon the amount of cyanogens (prussic acid) in the edible parts of the roots (Jones 1959, 12). The roots of sweet cassava are low in cyanogens, mealy after

cooking, and usually eaten as a raw vegetable, boiled, or roasted in an open fire. Bitter cassava varieties are high in cyanogens, waxy after cooking, and harmful to humans and animals unless they are peeled, grated, and toasted or soaked in water for a few days and boiled or sun-dried.

3. Some cities where this occurs are Dar es Salaam, Kampala, Kano, and Kinshasa.
4. This method was recently developed at the IITA and it is now widely used by farmers in the major cassava-producing countries in Africa.
5. Fifty percent of the farmer groups interviewed in the six COSCA study countries considered bright color as the most important quality in the cassava food products, namely the dried cassava roots, the granulated cassava food products, and the pasty cassava food products. Twenty-five percent considered texture, 15 percent considered taste, and 10 percent considered other qualities as the most important in the cassava food products.
6. Social scientists in the COSCA team interpreted this to mean that the wife's love was expressed in the pounding of cassava.
7. Cameroon is often considered to be in West or Central Africa.
8. It is also a common ingredient in pudding for desert in the United States.

## Chapter 9

1. The United Africa (trading) Company (UAC) was a colonial retail outlet.
2. Mechanized pressers are also used in oil-palm processing in some places.
3. Sam Duncan, alias Cassava Duncan, of Cape Coast (Doku 1969).
4. This section borrowed heavily from Idowu (1998)

## Chapter 10

1. Boserup (1970) pointed out that tree felling is nearly always done by men, but women are responsible for the removal and burning of the felled trees, planting, weeding, harvesting, processing, and food preparation.
2. The owner (either woman or man) of each cassava field was asked which gender provided the bulk of the labor for bush clearing, tillage, planting, weeding, harvesting, and transporting.
3. The ratio of men to women is low in female-headed households. A female is declared a household head if there is no adult male in the household.

Female-headed households are common in areas where men have migrated to mines, plantations, and urban centers.

4. The farmer groups interviewed in each village were asked who (men or women) mostly provided the bulk of the labor each cassava-processing task.

## Chapter 11

1. This chapter benefited from Ezemenari, Nweke, and Strauss 1998.
2. Waste was estimated to be 28 percent of the total cassava production in Africa from 1994 to 1998 (FAOSTAT).
3. The countries are: Angola, the Central African Republic, the Congo, the Peoples Republic of Congo, Ghana, Mozambique, and Nigeria.
4. The countries are: the Republic of Benin, Cameroon, the Côte d'Ivoire, Guinea, Liberia, Madagascar, Sierra Leone, Tanzania, Togo, Uganda, and Zambia.
5. The waste estimate of 28 percent for Africa in the late 1990s was influenced by the following figures: waste was 56 percent of annual cassava production in Nigeria, 35 percent in Ghana, 19 percent in Uganda, 12 percent in the Côte d'Ivoire, 11 percent in the Congo, and 2 percent in Tanzania (FAOSTAT). The FAO definition of cassava waste in Nigeria is "statistical difference between given (by Ministry of Agriculture) cassava production and given cassava consumption" (Tampieri 2001).
6. Household expenditure on cassava is used as a proxy for the amount of cassava consumed by the household.
7. Total household expenditure is used as a proxy for household income. The method for calculating the income elasticity of demand estimation is presented in appendix.

## Chapter 12

1. Personal interview, Abuja, 17 January 2001. Late in 2001, after the interview, the price of cassava rose sharply in Nigeria due to the increased demand for dried cassava roots for livestock feed in Europe following the outbreak of mad cow disease.
2. Between 1996 and 1998, there were 19.3 million cattle and 4.5 million pigs in Nigeria, compared with 156 million cattle and 28 million pigs in Brazil (FAO-

STAT). In Nigeria, nomadic herdsmen move their cattle to wherever grass is available and tsetse is not a problem. The nomads neither respect boundaries nor do they pay for grazing rights. Frequently, they are halted by crop farmers, including cassava farmers, when cattle graze on fields with crops.

3. In Nigeria, cassava leaves are returned to the soil as organic matter. Yet the cassava leaves are also a good source of protein and vitamins, and they are safe for human and animal consumption.
4. This section draws on Raw Materials Research and Development Council (RMRDC) 1997, 2000a, 2000b, and 2000c. There are gaps in the information contained in the RMRDC sources because a mail questionnaire was used by the RMRDC researchers to collect the data and some business organizations did not respond (RMRDC 2000a, 17).
5. The sources for this data are: Raw Materials Research and Development Council 1997, 65–85, 164, and 314–86; idem 2000a, 47; and idem 2000b, 67–75.
6. The sources for this data are: Raw Materials Research and Development Council 1997, 65–85, 164, and 314–86; idem 2000a, 47; idem 2000b, 67–75; and COSCA interviews of petroleum industry personnel, Port Harcourt, Nigeria, January 2001.
7. Personal interview, Onitsha, 13 January 2001.
8. In the late 1990s, however, the Nigerian textile industry operated at an average of only 35 percent of installed capacity (Raw Materials Research and Development Council 1996)
9. The Nigerian Starch Mill (NSM) has an established capacity of fifty tons of cassava starch per day. In early 2001, however, it was producing only about two tons per day.
10. Personal interview, Ihiala, 13 January 2001.
11. The sources for this data are: Raw Materials Research and Development Council 1997, 72–83 and idem 2000b, 71.
12. The sources for this data are Raw Materials Research and Development Council 1997, 83, 127, 314–19, and 385 and idem 2000b, 82.
13. The cost estimate for fresh root provided by Ogazi, Hassan, and Ogunwusi (1997, 77) was inflated by 25 percent to cover the cost of peeling and drying.
14. Personal interview, Umudike, 12 January 2001.

15. Personal interview, Umudike, 11 January 2001.
16. Personal interview, Onitsha, 12 January 2001.
17. The sources for this data are: Raw Materials Research and Development Council 1997, 71–84 and 317; idem 2000b, 67 and 77; and idem 2000c, 23–30.
18. One ton of fresh cassava roots yields 150 liters of alcohol (Balagopalan et al. 1988, 182)
19. Stillage is a distillery waste, which is produced in a volume of ten times that of ethanol. It can be used as fertilizer because it is rich in nitrogen, calcium, sulfate, and organic matter (Pimentel 1980).

# References

Adegboye, R. O., and J. A. Akinwumi. 1990. Cassava processing innovations in Nigeria. In *Tinker, tiller and technical change*, edited by H. Appleton and N. Carter, 64–79. London: Intermediate Technology Publications and New York: Bootstrap Press.

African Development Consulting Group (ADCG). 1997. Nigeria: Textile industry: Overview of supply and demand. Mbendi Information Services, *mbendi.co.za/adcg/txt103.htm.*

Akande, S. O. 2000. Price and trade policy. In Nigeria: Agricultural trade review (Draft). Washington, D.C: World Bank.

Akobundu, I. O. 1984. Effect of tillage on weed control in cassava. In *International Institute of Tropical Agriculture, Annual Report for 1993,* 163–64. Ibadan, Nigeria: International Institute of Tropical Agriculture.

Alderman, H. 1990. Nutritional status in Ghana and its determinants. Social dimensions of adjustment in sub-Saharan Africa. Policy Analysis. Working paper No. 3 Washington, D.C.: World Bank.

Al-Hassan, R. M. 1992. Industrial demand for cassava in Ghana: Prospects and problems. In *Tropical root crops: Promotion of root crop-based industries,* ed.

M. O. Akoroda and D. B. Arene, 269–73. Proceedings of the Fourth Triennial Symposium of the International Society for Tropical Root Crops-African Branch (ISTRC-AB), held in Kinshasa, Zaire, 5–8 December 1989. Ibadan, Nigeria: International Institute of Tropical Agriculture.

Almazan, M. 1987. *A guide to the picrate test for cyanide in cassava leaf.* Ibadan, Nigeria: International Institute of Tropical Agriculture.

Alternate Fuels Data Center (AFDC). 2000. Ethanol general information. *afdc.nrel.gov/altfuel/eth_general.htmlwhet*

Alyanak, Leyla. 1997. A modest grater means safer and quicker cassava in Uganda. *News Highlights. News/1997/970508-e.htm.* FAO Rome.

Anojulu, I. E. O. Adenola, and R. Ogunbiade. 1982. Review of NAFPP activities and its future role in Nigeria agricultural development. In *Proceedings of the National Agricultural Development Committee and 4th Joint National Accelerated Food Production Project Workshop, 17–20 February 1986, Moore Plantation, Ibadan,* 146–74. Lagos, Nigeria: Federal Department of Agriculture and Ibdan, Nigeria: National Cereals Research Institute.

Asadu, C. L. A., and Felix I. Nweke. 1999. Soils of arable crop fields in sub-Saharan Africa: Focus on cassava growing areas. Collaborative Study of Cassava in Africa (COSCA) Working Paper No. 18. Ibadan, Nigeria: International Institute of Tropical Agriculture.

Ay, Peter. 1991. Personal communication, 12 February.

Balagopalan, C., G. Padmaja, S. K. Nanda, and S. N. Moorthy, 1988. *Cassava in food, feed, and industry.* Boca Raton, Fla.: CRC Press.

Bamikole, O. T., and M. Bokanga. 2000. The ethanol industry in Nigeria growing on a new crop. Paper presented at the International Fuel Ethanol Workshop, Windsor, Ontario, Canada, 20–23 June 2000.

Beck, B. D. A. 1980. Historical perspectives of cassava breeding in Africa. In *Root crops in Eastern Africa,* 13–18. Proceedings of a Workshop held in Kigali, Rwanda, 23–27 November 1980. Ottawa, Canada: International Development Research Center (IDRC).

Berry, S. S. 1993. Socio-economic aspects of cassava cultivation and use in Africa: Implication for the development of appropriate technology. Collaborative Study of Cassava in Africa (COSCA) Working Paper No. 8. Ibadan, Nigeria: International Institute of Tropical Agriculture.

Blackie, M. J. 1990. Maize, food self-sufficiency and policy in East and Southern Africa. *Food Policy* 15:383–94.

Bokanga, M. 1992. Cassava fermentation and industrialization of cassava food production. In *Tropical root crops: Promotion of root crop-based industries*, ed. M. O. Akoroda and D. B. Arene, 197–201. Proceedings of the Fourth Triennial Symposium of the International Society for Tropical Root Crops-African Branch (ISTRC-AB), held in Kinshasa, Zaire, 5–8 December 1989. Ibadan, Nigeria: International Institute of Tropical Agriculture.

Bokanga, M. 2000. E-mail message. 22 September.

Bokanga, M., and O. O. Tewe. 1998. Cassava: A premium raw material for the food, feed, and industrial sectors in Africa. In *Post-harvest technology and commodity marketing: Proceedings of a post-harvest conference, 2 November to 1 December 1995, Accra, Ghana,* edited by R. S. B. Ferris. IITA: Ibadan.

Boserup, E. 1970. *Women's role in economic development*. New York: St. Martin's Press.

———. 1981. *Population and technological change: A study of long-term trends.* Chicago: University of Chicago Press.

Brock, J. E., 1955. Nutrition. *Annual Review of Biochemistry* 24:523–42.

Brown, L., H. Feldstein, L. Haddad, C. Pena, and A. Quisumbing. 1995. Generating food security in the year 2020: Women as producers, gatekeepers, and shock absorbers. Washington, D.C.: International Food Policy Research Institute.

Byerlee, D., and G. E. Alex, 1998. *Strengthening national agricultural research systems: Policy issues and good practice.* Washington, D.C.: World Bank.

Byerlee, D., and Carl K. Eicher, eds. 1997. *Africa's emerging maize revolution.* Boulder, Colo.: Lynne Rienner.

Camara, Youssouf. 2000. Profitability of cassava production systems in West Africa: A comparative analysis (Côte d'Ivoire, Ghana, and Nigeria). Ph.D. diss., Michigan State University, East Lansing.

Carter, S. E., and P. G. Jones. 1989. COSCA site selection procedure. Collaborative Study of Cassava in Africa (COSCA) Working Paper No. 2. Ibadan, Nigeria: International Institute of Tropical Agriculture.

Central Bureau of Statistics (CBS). 1979. *Summary report on household economic survey: 1974–1975.* Accra, Ghana: Central Bureau of Statistics.

Chiwona-Calton, L. 1995. Personal communication, 27 October.

Colonna, P., A. Buleon, and C. Mercier. 1987. Physically modified starches. In *Starch properties and potentials,* ed T. Galliad, 79–114. New York: John Wiley and Sons.

Cours, G., D. Forgette, G. W. Otim-Nape, and J. M. Thresh 1997. The epidemics of cassava mosaic virus disease in Madagascar in the 1930s–1940s: Lessons for the current situation in Uganda. *Tropical Science* 37:238–48.

Dahniya, M. T. 1983. Evaluation of cassava leaf and root production in Sierra Leone. In Proceedings: Sixth symposium of the International Society for Tropical Root Crops, 21–26 February, 299–302. International Potato Center, Lima, Peru.

Dahniya, M. T., and A. Jalloh. 1998. Relative effectiveness of sweet potato, melon and pumpkin as live-mulch in cassava. In *Root crops and poverty alleviation,* edited by M. O. Akoroda and I. J. Ekanayake, 353–55. Proceedings of the Sixth Triennial Symposium of the International Society for Tropical Root Crops - African Branch (ISTRC-AB), held in Lilongwe, Malawi, 22–28 October 1995. Ibadan, Nigeria: International Institute of Tropical Agriculture.

Davesne, A. 1950. *Manual d'agriculture.* Paris.

Deaton, A. 1997. *The analysis of household surveys: A microeconometric approach to development policy.* Baltimore, Md.: Johns Hopkins University Press.

Deaton, A., J. Ruiz-Castilo, and D. Thomas. 1989. The influence of household consumption on household expenditure patterns: Theory and Spanish evidence. *Journal of Political Economy* 97: 179–200.

Delfloor, I. 1995. Factors governing the bread-making potential of cassava flour in wheatless bread recipes. Ph.D. Dissertation. Kathnolieke University, Leuven.

Dixon, A. G. O. 2000. E-mail message, 10 October.

———. 2001. E-mail message, 9 February.

Dixon, A. G. O., R Asiedu, and S. K. Hahn. 1992. Cassava germplasm enhancement at the International Institute of Tropical Agriculture (IITA). In *Tropical root crops: Promotion of root crop-based industries. Proceedings of the fourth triennial symposium of the International Society for Tropical Root Crops–African Branch (ISTRC-AB),* ed. M. O. Akoroda and O. B. Arene, 83–87. Ibadan, Nigeria: International Institute of Tropical Agriculture.

Doku, E. V. 1966. Cultivated varieties of cassava in Ghana. *Ghana Journal of Science* 6:74–86.

———. 1969. *Cassava in Ghana.* Accra, Ghana: Ghana University Press.

———. 1989. Root crops in Ethiopia. In *Root crops and low input agriculture,* ed. M. N. Alvarez and S. K. Hahn. Proceedings of the Third East and Southern African Regional Root Crops Workshop, Mzuzu, Malawi, 7–11 December 1987. Ibadan, Nigeria: International Institute of Tropical Agriculture.

Drachoussoff, V., A. Focan, and J. Hecq. 1993. *Rural development in Central Africa: 1908 to 1960/1962.* Brussels: King Baudouin Foundation.

Eggleston, G., and P. Omoaka. 1994. Alternative breads from cassava flour. In *Tropical root crops in a developing economy: Proceedings of the ninth symposium of the International Society for Tropical Root Crops, 20–26 October 1991, Accra, Ghana,* edited by F. Ofori and S. K. Hahn, 243–48. IITA: Ibadan.

Eicher, C. K. 1982. Facing up to Africa's food crisis. *Foreign Affairs* 61 (1): 151–74.

Enete, A., F. I. Nweke, and E. C. Okorji. 1995. Trends in food crops yields under demographic pressure in sub-Saharan Africa: The case of cassava in southeast Nigeria. *Outlook on Agriculture* 24 (4): 249–54.

Ezemenari, K., F. I. Nweke, and J. Strauss. 1998. Consumption patterns and expenditure elasticities of demand for food staples in rural Africa: Focus on cassava growing area. Draft Collaborative Study of Cassava in Africa (COSCA) Working Paper. Ibadan, Nigeria: International Institute of Tropical Agriculture.

Ezumah, H. C., P. Ohuyon, and D. N. Kalabari. 1984. Effects of tillage, fertilizer, and mulch on cassava. In *International Institute of Tropical Agriculture annual report for 1993,* 164, 165. Ibadan, Nigeria: International Institute of Tropical Agriculture.

Ezumah, N. N., and C. M. D. Domenico. 1995. Enhancing the role of women in crop production: A case study of Igbo women in Nigeria. *World Development* 23 (10): 1731–44.

Federal Agricultural Coordinating Unit (FACU). 1986. Niger State Agricultural Development Project. Project Document, Multi-State Agricultural Development Project II, Phase I, 1988–1991. Kaduna, Nigeria: Federal Agriculture Coordinating Unit.

Food and Agriculture Organization of the United Nations (FAO). 1997. Raising women's productivity in agriculture. In *The state of food and agriculture 1997,* 58–71. Rome: Food and Agriculture Organization of the United Nations.

———. 1998. Women feed the world: FAO announces theme for World Food Day 1998. *News and Highlights.* News/1998/980305.

———. 2000. Women: Key to African development. News release No. 2000/340/AFR. Rome.

Fresco, L. 1986. *Cassava in shifting cultivation: A systems approach to agricultural technology development in Africa.* Amsterdam, The Netherlands: Royal Tropical Institute.

Fresco, Louise O. 1993. The dynamics of cassava in Africa: An outline of research issues. Collaborative Study of Cassava in Africa (COSCA) Working Paper No. 9. Ibadan, Nigeria: International Institute of Tropical Agriculture.

Garman, C., and N. C. Navasero. 1982. Evaluation of harvesting machines and tools. In *International Institute of Tropical Agriculture, Annual Report for 1981,* 40–41. Ibadan, Nigeria: International Institute of Tropical Agriculture.

Gladwin, C. H. 1996. Gender in research design: Old debates and new issues. In *Achieving greater impact from research investments in Africa,* ed. S. A. Breth, 127–49. Mexico City: Sasakawa Africa Association.

Grace, M. R. 1977. Cassava processing. FAO Plant Production and Protection Series No. 3. Rome: Food and Agriculture Organization of the United Nations.

GreenTimes. 2000. Proalcool: The Brazilian alcohol programme. Greentie. *www.greentie.org.*

Hahn, N. D. 1985 Socioeconomic assessments. In *International Institute of Tropical Agriculture, Annual Report for 1985,* 198–200. Ibadan, Nigeria: International Institute of Tropical Agriculture.

Hahn, S. K. 1984. Tropical root crops: Their improvement and utilization. Conference Paper 2. Ibadan, Nigeria: International Institute of Tropical Agriculture

———. 1989. An overview of African traditional cassava processing and utilization. *Outlook on Agriculture* 18 (3): 110–18.

———. 1999. Text of seminar presented at the African Studies Center, Michigan

State University, East Lansing, Michigan, 28 January.

———. 2000. Personal communication, 17 October.

Hahn, S. K., A. K. Howland, and E. R. Terry. 1980. Correlated resistance of cassava to mosaic and bacterial blight diseases. *Euphytica* 29:305–11.

Hahn, S. K., E. R. Terry, K. Leuschner, and T. P. Singh. 1981. Cassava improvement strategies for resistance to major economic diseases and pests in Africa. In *Tropical root crops research strategies for the 1980s,* ed. E. R. Terry, K. O. Oduro, and F. Caveness, 25–28. Proceedings of the First Triennial Symposium of the International Society for Tropical Root Crops–African Branch (ISTRC-AB), 8–12 September 1980. Ottawa, Canada: International Development Research Center.

Harrison, B. D., Y. L. Liu, X. Zhou, D. J. Robinson, L. Calvert, C. Munoz, and G. W. Otem-Nape. 1997. Properties, differentiation and geographical distribution of geminiviruses that cause cassava mosaic disease. *African Journal of Root and Tuber Crops* 2 (1 and 2): 19–22. International Society for Tropical Root Crops–African Branch (ISTRC-AB). Ibadan, Nigeria.

Hendershott, C. H. ed. 1972. A Literature Review and Research Recommendations on Cassava, Manihot esculenta, Crantz. Unpublished paper. University of Georgia, Athens, Georgia

Herren, H. R. 1981. Biological control of the cassava mealybug. In *Tropical root crops research strategies for the 1980s,* ed. E. R. Terry, K. O. Oduro, and F. Caveness, 79, 80. Proceedings of the First Triennial Symposium of the International Society for Tropical Root Crops–African Branch, 8–12 September 1980. Ottawa, Canada: International Development Research Center, Ottawa, Canada.

Herren, H. R., P. Neuenschwander, R. D. Hennessey, and W. N. O. Hammond. 1987. Introduction and dispersal of *Epidinocarisi lopezi* (Hym. Encytidae), an exotic parastoid of cassava mealybug, *Phenococcus manihoti* (Hom. Pseudococcidae) in Africa. *Agriculture, Ecosystems, and Environments* 19:131–44.

Heys, G. 1977. Cassava improvement in the Niger Delta of Nigeria. In *Cassava bacterial blight,* ed. G. Persley, E. R. Terry, and R. MacIntyre, 18, 19. Report of an Interdisciplinary Workshop held at IITA, Ibadan, Nigeria, 1–4 November 1976. Ibadan, Nigeria: International Institute of Tropical Agriculture.

Hirschman, A. 1981. The rise and decline of development economics. In *Essays in trespassing: Economics to politics and beyond,* ed. A. O. Hirschman, ed. New York: Cambridge University Press.

Hotelling, H. 1933. Analysis of complex statistical variables into principal components. *Journal of Educational Psychology* 24: 417–41 and 498–520.

Idachaba, F. S. 1998. Instability of national agricultural research systems in sub-Saharan Africa: Lessons from Nigeria. Research Report No. 13, International Service for National Agricultural Research (ISNAR), The Hague.

Idowu, I. A. 1998. Private sector participation in agricultural research and technology transfer linkages: Lessons from cassava *gari* processing technology in Southern Nigeria. In *Post-harvest technology and commodity marketing,* ed. R. S. B. Ferris, 85–90. Proceedings of a Post-Harvest Conference, 2 November to 1 December 1995, Accra, Ghana. Ibadan, Nigeria: International Institute of Tropical Agriculture.

International Food Policy Research Institute (IFPRI). 1976. Meeting food needs in the developing world: The location and magnitude of the task in the next decade. Research Report No. 1. International Food Policy Research Institute, Washington, D.C.

International Fund for Agricultural Development (IFAD) and Food and Agriculture Organization of the United Nations (FAO). 2000. *The world cassava economy: Facts and outlook*. Rome: IFAD.

International Institute of Tropical Agriculture (IITA). 1986. *Root and tuber improvement program annual report for 1985*. Ibadan, Nigeria: International Institute of Tropical Agriculture.

———. 1992a. Plant health management. In *Sustainable food production in sub-Saharan Africa: 1. IITA's contributions*, 139–69. Ibadan, Nigeria: International Institute of Tropical Agriculture.

———. 1992b. Resource and crop management. In *Sustainable food production in sub-Saharan Africa: 1. IITA's contributions,* 25–63. Ibadan, Nigeria: International Institute of Tropical Agriculture.

———. 1994. Cassava's expanding prospects. In *International Institute of Tropical Agriculture Annual report 1994,* 36–43. Ibadan, Nigeria: International Institute of Tropical Agriculture.

———. 1995. Cassava in Malawi: From drought to food security. In *International*

*Institute of Tropical Agriculture annual report 1995*, 31–34. Ibadan, Nigeria: International Institute of Tropical Agriculture.

———. 1996. Biocontrol of cassava green mite gives African farmers a bonanza. In *International Institute of Tropical Agriculture annual report 1996*, 31–32. Ibadan, Nigeria: International Institute of Tropical Agriculture.

———. 1998. Restoring nutrients to sub-Saharan soils. In *International Institute of Tropical Agriculture annual report 1997*, 18–19. Ibadan, Nigeria: International Institute of Tropical Agriculture.

Ikpi, A. E. 1989a. Cassava development in Nigeria. In *Cassava: Lifeline for the rural households*, ed. A. E. Ikpi and N. D. Hahn, 1–15. Lagos, Nigeria: UNICEF.

———. 1989b. Economic considerations of cassava development. In *Cassava: Lifeline for the rural households*, ed. A. E. Ikpi and N. D. Hahn, 117–25. Lagos, Nigeria: UNICEF.

Ikpi, A. E., and N. D. Hahn, eds. 1989. Cassava: Lifeline for the rural households. Lagos, Nigeria: UNICEF.

Irvine, F. R. 1953. A text book of West African agriculture, soils, and crops. London: Oxford University Press.

Jennings, D. L. 1976. Breeding for resistance to African cassava mosaic disease: Progress and prospects. In *African cassava mosaic*, ed. B. L. Nestel, 3 9–44. Report of an interdisciplinary workshop held at Muguga, Kenya, 19–22 February 1976. Ottawa, Canada: International Development Research Center.

Johnson, D. G. 2000. Population, food, and knowledge. *American Economic Review* 90 (1): 1–14.

Johnston, B. F. 1958. *The staple food economies of western tropical Africa*. Stanford, Calif.: Stanford University Press.

Jones, W. O. 1959. *Manioc in Africa*. Stanford, Calif.: Food Research Institute, Stanford University.

———. 1972. *Marketing staple foods in tropical Africa*. Ithaca: Cornell University Press.

Kapinga, R. 1995. Personal communication, 27 October.

Lagemann, J. 1977. *Traditional African farming systems in eastern Nigeria*. African Studies 98. Munich: Weltforum Verlag.

Lahai, B. A. N., P. Goldey, and G. E. Jones. 2000. The gender of the extension agent and farmers' access to and participation in agricultural extension in Nigeria. *Journal of Agricultural Education and Extension* 6 (4): 223–33.

Latham, M. C. 1979. *Human nutrition in tropical Africa.* Rome: Food and Agriculture Organization of the United Nations.

Legg, J. 1998. Cassava mosaic pandemic threatens food security. *AgriForum: Quarterly News Letter of the Association for Strengthening Agricultural Research in Eastern and Central Africa* 3:1–8.

Leser, C. E. V. 1963. Forms of Engel functions. *Econometrica* 31:694–703.

Lindeman, L. R., and C. Rocchiccioli. 1979. Ethanol in Brazil: Brief summary of the state of industry in 1977. *Biotechnology and Bioengineering* 21 (7): 1107–19.

Lutaladio, N. B., and H. C. Ezumah. n.d. Cassava leaf harvesting in Zaire (Congo). M'vuazi.Unpublished memorandum., research station in the Congo.

Makambila, Casimir. 1981. Cassava root rot due to *Amillariella tabescens* in the People's Republic of Congo. In *Tropical root crops research strategies for the 1980s*, ed. E. R. Terry, K. O. Oduro, and F. Caveness, 69–74. Proceedings of the First Triennial Root Crops Symposium of the International Society for Tropical Root Crops–African Branch (ISTRC-AB), 8–12 September 1980, Ibadan. Ottawa, Canada: International Development Research Center.

Martin, S. 1984. Gender and innovation: Farming, cooking, and palm processing in the Ngwa Region, southeastern Nigeria, 1900–1930. *Journal of African History* 25: 411–27.

de Matos, Peres Aristoteles. 2001. E-mail message. 16 January.

Mba, C. 2000. E-mail message. 21 September.

Mba, R. E. C., and A. G. O. Dixon. 1998. Genotype × environment interaction, phenotipic stability of cassava yields and heritability estimates for production and pest resistance traits in Nigeria. In *Root crops and poverty alleviation,* ed. M. O. Akorada and I. J. Ikanayake, 255–61. Proceedings of the Sixth Triennial Symposium of the International Society for Tropical Root Crops-African Branch (ISTRC-AB), Lilongwe, Malawi, 22–28 October 1995. Ibadan, Nigeria: International Institute of Tropical Agriculture.

Mbendi. 2000. Nigeria: Oil and gas industry. *mbendi.co.za/indy/oilg/ng*

McCalla, A. F. 1999. Prospects for food security in the twenty-first century: With special emphasis on Africa. *Agricultural Economics* 20:95–103.

Mellor, J. W., C. Delgado, and M. Blackie, eds. 1987. *Accelerating food production in sub-Saharan Africa.* Baltimore, Md.: Johns Hopkins University Press.

Mkumbira, J. 1998. Multiplication and distribution of cassava and sweet potato planting materials: Lessons from Malawi and implications for the region. In *Root crops and poverty alleviation,* ed. M. O. Akorada and I. J. Ikanayake, 51–55. Proceedings of the Sixth Triennial Symposium of the International Society for Tropical Root Crops–African Branch (ISTRC-AB), Lilongwe, Malawi, 22–28 October 1995. Ibadan, Nigeria: International Institute of Tropical Agriculture.

Msabaha, M. A. M., and B. W. Rwenyagira. 1989. Cassava production, consumption, and research in Tanzania. In *Status of data on cassava in major producing countries in Africa,* ed. F. I. Nweke, J. Lynam, and C. Prudencio, 22–27. Collaborative Study of Cassava in Africa (COSCA) Working Paper No. 3. Ibadan, Nigeria: IITA.

Nartey, F. 1977. *Manihot esculenta* (cassava): Cyanogenesis, ultra-structure and seed germination. Ph.D. diss., University of Copenhagen, Munksgaard, Copenhagen, Denmark.

Ndamage, G. 1991. Personal communication, 26 October.

Neuenschwander, P., W. N. O. Hammond, O. Ajunonu, A. Gabo, T. N. Echendu, A. H. Bokonon-Ganta, R. Allomasso, and I. Okon. 1990. Biological control of cassava mealybug, *Phenacoccus manihot* (Hom., Pseudococcidae) by *Epidinocarsis lopezi* (Hym., Encyrtidae) in West Africa, as influenced by climate and soil. *Agriculture, Ecosystems and Environment* 32:39–55.

Ng, N. Q. 1992. Unpublished data. Genetic Resources Unit. Ibadan, Nigeria: IITA.

Ngoddy, P. O. 1977. Determinants of the development of technology for processing of roots and tubers in Nigeria. *Proceedings of the First National Seminar on Root and Tuber Crops.* Umudike, Umuahia, Nigeria.: National Root Crops Research Institute.

Nichols, R. F. W. 1947. Breeding cassava for virus resistance. *East African Agricultural Journal* 12:184–94.

Nicol, B. M. 1954. A report of the nutritional work which has been carried out in Nigeria since 1920, with a summary of what is known of the present nutritional state of the Nigerian peasants Nigeria, Eastern Region, Ministry of Health. Mimeograph.

Nnanyelugo, D. O., E. K. Ngwu, C. O. Asinobi, A. C. Uwaegbute, and E. C. Okeke. 1992. *Nutrient requirements and nutritional status of Nigerians including policy considerations.* Ibadan, Nigeria: African Book Builders.

Norman, D. W. 1974. Rationalizing mixed cropping under indigenous conditions: The example of Northern Nigeria. *Journal of Development Studies* 11:3–21.

Nweke, F. I. 1992. Economics of root crops processing and utilization in Africa: A challenge for research. In *Tropical root crops: Promotion of root crop-based industries,* ed. M. O. Akoroda and O. B. Arene, 27–36. Proceedings of the Fourth Triennial Symposium of the International Society for Tropical Root Crops-African Branch (ISTRC-AB). Ibadan, Nigeria: International Institute of Tropical Agriculture.

Nweke, F. I., E. C. Okorji, J. E. Njoku, and D. J. King. 1994. Expenditure elasticities of demand for major food items in south-east Nigeria. *Tropical Agriculture* (Trinidad) 71 (3): 229–34.

Nye, P. H., and D. J. Greenland. 1960. The soil under shifting cultivation. Herpenden, England, Commonwealth Bureau of Soils, Tech. Comm. 51.

Odurukwe, S. O., and U. I. Oji. 1981. The effects of previous cropping on yields of yam, cassava and maize. In *Tropical root crops: Research strategies for the 1980s,* 116–19. Proceedings of the First Triennial Symposium of the International Society for Tropical Root Crops–African Branch (ISTRC-AB), 8–12 September 1980, Ibadan, Nigeria. Ottawa, Canada: International Development Research Center.

Ogazi, Paul O., Usman A. Hassan, and Abimbola A. Ogunwusi. 1997. Boosting the supply of agricultural raw materials for industrial use: The RMRDC experience. Abuja, Nigeria: Raw Materials Research and Development Council.

Ohunyon, P. U., and J. A. Ogio-Okirika. 1979. Eradication of cassava bacterial blight/cassava improvement in the Niger Delta of Nigeria. In *Cassava bacterial blight in Africa: Past, present, and future,* ed. E. R. Terry, G. J. Persley, and S. C. A. Cook, 55–56. Report of an Interdisciplinary Workshop held at IITA, Ibadan, Nigeria, 26–30 June 1978. Ibadan, Nigeria: International Institute of Tropical Agriculture.

Okigbo, Bede N. 1980. Nutritional implications of projects giving high priority to the production of staples of low nutritive quality. The case of cassava in the humid tropics of West Africa. *Food and Nutrition Bulletin,* United Nations

University, Tokyo, 2 (4): 1–10.

———. 1984. Improved permanent production systems as alternative to shifting intermittent cultivation. In Improved production systems as an alternative to shifting cultivation. *FAO Soils Bulletin* 53:1–100. Soils Resources Management and Conservation Service, Land and Water Development Division, Food and Agriculture Organization of the United Nations, Rome, Italy.

———. 1998. Personal communication, 1 March.

Okoli, O. O. 2000. E-mail message. 29 March.

Okorji, E. C. 1983. Consequences for agricultural productivity of crop stereotyping along sex lines: A case study of four villages in Abakaliki area of Anambra state. Master's thesis, Department of Agricultural Economics, University of Nigeria, Nsukka, Nigeria.

Omawale, and A. M. Rodrigues. 1980. Nutrition considerations in a cassava production program for Guyana. *Ecology of Food and Nutrition* 10:87–95. New York, London, Paris: Gordon and Breach Science Publishers.

Onabolu, A. 2001. Cassava processing, consumption and dietary cyanide exposure, Ph.D. diss. Karolinska Institutet, Sweden.

Onabolu, A., A. Abass, and M. Bokanga. 1998. *New food products from cassava.* Ibadan, Nigeria: International Institute of Tropical Agriculture.

Onwueme, I. C. 1978. *The tropical tuber crops.* New York: John Wiley and Sons.

Onwueme, I. C., and T. D. Sinha. 1991. Field crop production in Tropical Africa. Technical Centre for Agriculture and Rural Cooperation (CTA), Ede, the Netherlands.

Onwuta, V. 2001. Personal communication, 9 January.

Osiname, A. O. , C. Bartlett, N. Mbulu, L. Simba, and K. Landu. 1988. Diagnostic survey of cassava-based cropping systems in two ecological zones of Bas-Zaire (Congo). In *Linking similar environments,* 73–82. Contributions from the First Annual Meeting of the Collaborative Group in Cassava-based Cropping Systems Research. Cassava-Based Cropping Systems Research 1. Resource and Crops Management Program. Ibadan, Nigeria: International Institute of Tropical Agriculture.

Osuntokun, B. O. 1973. Ataxic neuropathy associated with high cassava diets in West Africa. In *Chronic cassava toxicity,* ed. B. Nestel and R. MacIntyre, 127–38. Proceedings of an Interdisciplinary Workshop held in London,

29–30 January 1973. Ottawa, Canada: International Development Research Center.

———. 1981. Cassava diet, chronic cyanide intoxication and neuropathy in the Nigerian Africans. *World Review of Nutrition and Diet* 36:141–73.

Oti, E., and G. Asumugha. 2001. Personal communication. 11 January.

Otim-Nape, G. W. 1995. Personal communication, 27 October.

Otim-Nape, G. W., A. Bua, J. M. Thresh, Y. Baguma, S. Ogwal, G. N. Ssemakula, G. Acola, B. Byabakama, J. B. Colvin, R. J. Cooter, and A. Martin. 2000. *The current pandemic of cassava mosaic virus disease in East Africa and its control.* Chatham, U.K.: Natural Resources Institute.

Otim-Nape, G. W., A. Bua, J. M. Thresh, Y. Baguma, S. Ogwal, G. N. Ssemakula, G. Acola, B. Byabakama, and A. Martin. 1997. *Cassava mosaic virus disease in Uganda: The current pandemic of and approaches to control.* Chatham, U.K.: Natural Resources Institute.

Otim-Nape, G. W., and J. U. A. Opio-Odongo. 1989. Cassava in Uganda. In *Status of data on cassava in major producing countries in Africa,* ed. F. I. Nweke, J. Lynam, and C. Prudencio, 28–32. Collaborative Study of Cassava in Africa (COSCA) Working Paper No. 3. Ibadan, Nigeria: International Institute of Tropical Agriculture.

Otoo, J. 2000. E-mail message. 1 November.

Otoo, J., and J. J. Afuakwa. 2001. E-mail message, 27 February.

Otoo, J. A., A. G. O. Dixon, R. Asiedu, J. E. Okeke, G. N. Maroya, K. Tougnon, O. O. Okoli, J. P. Tette, and S. K. Hahn. 1994. Genotype x environment interaction studies with cassava. In *Tropical root crops in developing countries,* ed. F. Ofori and S. K. Hahn, 146–48. Proceedings of the International Society for Tropical Root Crops, 20 to 26 October 1991, Accra, Ghana.

Palmer-Jones, I. 1991. Gender and population in the adjustment of African economies: Planning for change. Women, Work and Development No. 19, International Labor Office, Geneva, Switzerland.

Pearson, K. 1901. On lines and planes closest fit to systems of points in space. *Philosophical Magazine* 6 (2): 559–72.

Persley, G. J. 1977. Distribution and importance of cassava bacterial blight in Africa. In *Cassava bacterial blight,* ed. G. Persley, E. R. Terry, and R. MacIntyre, 9–14. Report of an Interdisciplinary Workshop held at IITA, Ibadan,

Nigeria, 1–4 November 1976. Ibadan, Nigeria: International Institute of Tropical Agriculture.

Phillips, T. P. 1973. *Cassava utilization and potential markets.* Ottawa, Canada: International Development Research Center.

Pimentel, L. S. 1980. Biotechnology report: The Brazilian ethanol program. *Biotechnology and Bioengineering* 22 (10): 1989–2001.

Pingali, P., Y. Bigot, and H. P. Binswanger. 1987. *Agricultural mechanization and the evolution of farming systems in sub-Saharan Africa.* Washington, D.C.: World Bank.

Pinstrup-Anderson, P., P. Pandya-Lorch, and M. W. Rosegrant. 1999. World food prospects: Critical issues for the early twenty-first century. 2020 Vision, Food Policy Report. Washington, D.C.: International Food Policy Research Institute.

Platt, B. S. 1954. Food and its production in relation to the development of tropical and sub-tropical countries. In *The development of tropical and sub-topical countries,* edited by A. L. Banks, 97–115 and 136–37. London: Edward Arnold Ltd.

Quisumbing, A. 1996. Male-female difference in agricultural productivity. *World Development* 24:1579–95.

Ratanawaraha, C., N. Senanarong, and P. Suriyapan. 1999. Status of cassava in Thailand: Implications for future research and development. Global Cassava Development Strategy Validation Forum. Rome: Food and Agriculture Organization of the United Nations.

Raw Materials Research and Development Council (RMRDC). 1996. *Update on techno-economic survey of the multi-disciplinary task force on textile, weaving apparel, and leather sector.* Abuja, Nigeria: RMRDC.

———. 1997. *Raw materials sourcing for manufacturing in Nigeria.* 3d ed. Abuja, Nigeria: RMRDC.

———. 2000a. Base metal, iron and steel, and engineering services sector. Multi-disciplinary task force report of techno-economic survey. 3d ed. Abuja, Nigeria: RMRDC.

———. 2000b. Chemicals and pharmaceuticals sector. Multi-disciplinary task force report of techno-economic survey. 3d ed. Abuja, Nigeria: RMRDC.

———. 2000c. Food, beverage, and tobacco sector. Multi-disciplinary task force report of techno-economic survey. 3d ed. Abuja, Nigeria: RMRDC.

Rosling, Hans. 1986. Cassava, cyanide and epidemic spastic paraparesis: A study in Mozambique on dietary cyanide exposure. ACTA Universitatis Upsalienis, Ph.D. diss., Uppsala University, Uppasala, Sweden.

Rossel, H. W., C. M. Changa, and G. I. Atiri. 1994. Quantification of resistance to African cassava mosaic virus (ACMV) in IITA improved mosaic-resistant cassava breeding materials. In *Food crops for food security in Africa,* ed. M. O. Akoroda, 280–87. Proceedings of the Fifth Triennial Symposium of the International Society of Tropical Root Crops–African Branch (ISTRC-AB), 22–28 November 1992. Ibadan, Nigeria: International Society for Tropical Root Crops/The Technical Center for Agricultural and Rural Cooperation/International Institute for Tropical Agriculture.

Ruttan, V. 2001. *Technology, growth, and development: An induced innovation perspective.* New York: Oxford University Press.

Satin, M.. n.d.. *Functional properties of starches.* Rome: Food and Agriculture Organization of the United Nations.

———. 1988. *Bread without wheat. New Scientist* (28 April): 56–59.

Schaab, R. P., J. Zeddies, P. Neuenschwander, and H. Herren. 1998. Economics of biological control of the cassava mealybug *Phenacoccus manihoti* (Matt.-Ferr.) (Hom., Pseudococcidae) in Africa. In *Root crops and poverty alleviation,* 319–27. Proceedings of the Sixth Triennial Symposium of the International Society of Tropical Root Crops–African Branch (ISTRC-AB), 22–28 October 1995, Lilongwe, Malawi.

Scott, G. J., M. W. Rosegrant, and C. Ringler. 2000. Roots and tubers for the 21st century: Trends, projections, and policy options. Food, Agriculture, and Environment Discussion Paper 31, International Food Policy Research Institute (IFPRI), Washington, D.C.

Snyder, Margaret. 1990. Women: Key to ending hunger. The Hunger Project Papers, No. 8. New York: Global Hunger Project.

Southern African Development Coordination Conference (SADCC). 1984. Drought. Proceedings of conference held in Lusaka, Republic of Zambia, 2–3 February 1984.

Spencer, D. S. C. 1976. African women in agricultural development: A case study in Sierra Leone. American Overseas Liaison Committee Paper No. 9. Washington, D.C.: American Council on Education.

Stoekli, B. 1998. Bioplastics from transgenic cassava: A pipe dream, or a promising subject for research cooperation. *Agriculture and Rural Development* 5(2): 57–59.

Storey, H. H., and R. F. W. Nichols. 1938. Studies of the mosaic diseases of cassava. *Annals of Applied Biology.* 25 (4): 790–806.

Tampieri, O. 2001. E-mail message, 14 March.

Tanzania Food and Nutrition Center (TFNC) and International Child Health Unit (ICHU). n.d. Project proposal: cassava processing in Tanzania with special reference to cyanide toxicity. TFNC, Dar es Salaam, Tanzania, and ICHU, University Hospital, Uppsala, Sweden.

Tata-Hangy, K. 2000. Cassava green mite fighter faces problems. *AgriForum* 12:5, 6. The Association for Strengthening Agricultural Research in Eastern and Central Africa (ASARECA).

Terry, E. R. 1979. Diagnosis of bacterial blight disease. In *Cassava bacterial blight in Africa: Past, present, and future,* ed. E. R. Terry, G. J. Persley, and S. C. A. Cook, 5–8. Report of an Interdisciplinary Workshop held at IITA, Ibadan, Nigeria, 26–30 June 1978. Ibadan, Nigeria: International Institute of Tropical Agriculture

Tewe, O. O., and M. Bokanga. 2001. Unpublished data. Research highlights: Cassava utilization, 2000. Ibadan, Nigeria: International Institute of Tropical Agriculture.

Thomas, D., J. Strauss, and M. M. T. L. Barbosa. 1991. Estimativas do impacto de mudancas de renda e de precos no consumo no Brasil. Pesquisa e planejamento economico. *Instituto de Pes quisa Economica Applicada* 21 (2): 305–54.

Thresh, J. M., G. W. Otim-Nape, J. P. Legg, and D. Fargette. 1997. African cassava mosaic virus disease: The magnitude of the problem. *African Journal of Root and Tuber Crops* 2 (1 and 2): 13–19. The International Society of Tropical Root Crops-African Branch (ISTRC-AB), IITA, Ibadan, Nigeria.

Thro, A. M. 1998. Cassava biotec network, update to the ISTRC-AB: Network activities and scientific progress. In *Root crops and poverty alleviation,* ed. M. O. Akoroda and I. J. Ekanayake, 404–8. Proceedings of the Sixth Triennial Symposium of the International Society for Tropical Root Crops - African Branch (ISTRC-AB), held in Lilongwe, Malawi, 22–28 October 1995.

Ibadan, Nigeria: International Institute of Tropical Agriculture.

Tollens, E. 1992. Cassava marketing in Zaire—An analysis of its structure, conduct, and performance. Katholieke Universiteit Leuven, Belgium. Mimeograph.

Udry, C., J. Hoddinott, H. Alderman, and L. Haddad. 1995. Gender differentials in farm productivity: Implications for household efficiency and agricultural policy. *Food Policy* 20:407–23.

United States Aid for International Development (USAID). 1997. Women: An untapped resource for agricultural growth. *Genderaction: A newsletter of the USAID office of women in development* (Washington, D.C.) 1 (3): 2–5.

White, L. 1990. *Magomero: Portrait of an African village.* Cambridge: Cambridge University Press.

Wigg, D. 1993. *The quiet revolutionaries: A look at the campaign by agricultural scientists to fight hunger.* World Bank Development Essays No. 2. World Bank.

Working, H. 1943. Statistical laws of family expenditure. *Journal of the American Statistical Association* 38:43–56.

World Bank. 1990. *Women in development: A progress report on the World Bank Initiative.* Washington, D.C.: World Bank.

———. 1994. *Enhancing women's participation in economic development.* A World Bank Policy Paper. Washington, D.C.: World Bank,.

———. 1999. *World development report, 1999/2000.* Washington, D.C.: World Bank.

———. 2000. Women: Key to African development. News Release No: 2000/340/AFR. Washington, D.C.: World Bank.

———. 2001. Implementation completion report (2446-Uganda) on credit in the amount of US$28.8 million to the Republic of Uganda for an agricultural research and training (February 2001). Washington, D.C.

———, ed. 1993. Performance audit report: Nigeria: Bauchi State Agricultural Development Project; Kano State Agricultural Development Project; Sokoto State Agricultural Development Project; Ilorin Agricultural Development Project; Oyo North Agricultural Development Project; Agricultural Technical Assistance Project. Washington, D.C..: Operations Evaluation Department, The Word Bank.

Vilpoux, O., and M. T. Ospina. 1999. Cassava foods, global opportunities, case study: Brazil. In *Global cassava market study: Business opportunities for the use of cassava (dTp Studies Inc.).* The Global Development Cassava Strategy Validation Forum, 28 April 2000. Rome: FAO.

Yaninek, J. S. 1994. Cassava plant protection in Africa. In *Food crops for food security in Africa,* ed. M. O. Akoroda, 26–37. Proceedings of the Fifth Triennial Symposium of the International Society for Tropical Root Crops–African Branch (ISTRC-AB). Ibadan, Nigeria: International Society for Tropical Root Crops/The Technical Center for Agricultural and Rural Cooperation/International Institute for Tropical Agriculture.

# About the Authors

### Felix I. Nweke

Felix I. Nweke is a Visiting Professor of Agricultural Economics and African Studies at Michigan State University. He received his undergraduate degree in Agricultural Economics at the University of Nigeria, Nsukka, and his Ph.D. at Michigan State University. He then joined the International Food Policy Research Institute (IFPRI) and carried out a study of food policy in Ghana. Dr. Nweke joined the University of Nigeria in 1977 and taught agricultural economics for a decade. He also carried out in-depth studies of root crops in Nigeria from 1977 to 1987. In 1987, Dr. Nweke joined the International Institute of Tropical Agriculture (IITA) and served as the leader of the Collaborative Study of Cassava in Africa (COSCA) study. In 1995, he was rewarded with the Distinguished Service Award of the International Society for Tropical Root Crops–African Branch (ISTRC-AB) for outstanding leadership in the COSCA study.

### Dunstan S. C. Spencer

Dunstan S. C. Spencer is an independent consultant, based in Freetown, Sierra Leone, where he has been a team member or leader in a number of agricultural project identification, design, appraisal, supervision, and evaluation missions on behalf of several international development and donor agencies. Dr. Spencer has taught agricultural and development economics to undergraduate students in Sierra Leone and graduate students in the United States. He has led major farming systems research projects in International Agricultural Research Centers (IARCs) including the IITA, where he was the Director of the Resource and Crop Management Division which provided administrative support for the COSCA study.

### John K. Lynam

John K. Lynam is Associate Director of Agricultural Programs in the Rockefeller Foundation. He received his Ph.D. degree from the Food Research Institute at Stanford University, when William O. Jones was a faculty member carrying out research on food crops in Africa. Dr. Lynam spent eleven years as an economist in the Cassava Program in the International Center for Tropical Agriculture (CIAT) in Cali, Colombia. He was in charge of economic research program on cassava in Latin America and Asia. For the past twelve years, Dr. Lynam has been in charge of agricultural programs for the Rockefeller Foundation in East Africa. He played a major role in designing the COSCA study.

# Index

## D

## G



## H



## I

**J**



**K**



**L**

**M**



**N**

## O



## P



## R

**U**



**V**



**W**